AF454334

# Achieving Zero Hunger

## THE EDITORS

 **Dr. Zahoor A Pampori,** Professor and Head, Division of Veterinary Physiology, Faculty of Veterinary Sciences, SKUAST-Kashmir, Srinagar. Has graduated in veterinary sciences from Ranchi Veterinary College in 1989 and did doctorate in Animal Physiology from National Dairy Research Institute, Karnal, Haryana. The editor has served 13 years in State Animal Husbandry Department and is discharging his duties in the Faculty of Veterinary Sciences, SKUAST-K for last 20 years in different capacities. He has over 40 publications in international and national journals, authored three books and two manuals. He has organized several national level training programmes and workshops.

# Achieving Zero Hunger

**Zahoor A. Pampori**

*Faculty of Veterinary Sciences and AH,
SKUAST-Kashmir,
Srinagar, J&K*

2022

Daya Publishing House®
*A Division of*

Astral International Pvt. Ltd.
New Delhi – 110 002

*Published by* : **Daya Publishing House**®
*A Division of*
**Astral International Pvt. Ltd.**
– ISO 9001:2015 Certified Company –
4736/23, Ansari Road, Darya Ganj,
New Delhi-110 002
Ph. 011-43549197, 8130496929
E-mail: info@astralint.com
Website: www.astralint.com

# Acknowledgement

I, Z.A. Pampori, the editor of the book wish to put on record the moral and financial support rendered by the Principal Investigator, National Agriculture Higher Education Project – SKUAST-Kashmir, Prof. Nazir Ahmad Ganai. The editor sincerely & gratefully acknowledges the logistic and financial support provided by the NAHEP- SKUAST-K, without which it was not possible to bring out this publication. The book was carved out of the lectures delivered by reputed national and international experts in a workshop on "Achieving zero hunger by 2030: the critical role played by agriculture and allied sectors" that was organised by the Faculty of Veterinary Sciences & Animal Husbandry, SKUAST-K and sponsored by NAHEP-ICAR. Therefore the book is in real sense dedicated to the NAHEP-SKUAST-Kashmir.

*Zahoor A. Pampori*

# Foreword

The world population has increased from 1 billion in 1800AD to 7.9 billion in 2020AD. However, a significant increase in the human population has been witnessed post industrialisation since the 1950s, mainly due to medical advancements and a rise in agricultural productivity, which raised human life expectancy. Subsequent to the increased demand for food, the production per unit of land has also increased several folds due to genetically improved varieties, mechanised agriculture practices, and fertilizers. However, with intensive cultivation, grazing, soil mineralisation, overuse of chemicals, deforestation, climate change and environmental degradation, sustaining the food production for the growing global population is a challenge.

Around 815 million people in the world, mostly from underdeveloped countries, face hunger, out of which about 135 million suffer from acute hunger. Child mortality, stunting and malnutrition are the consequences of food shortage.

In 2015, all the United Nations member states adopted a universal call to action to end poverty and hunger, protect the planet and ensure that all people enjoy peace and prosperity by 2030, commonly referred to as the "Sustainable Development Goals" (SDGs). Achieving zero hunger in the world is one of the SDGs to end hunger and its all forms by 2030. Unfortunately, the Covid -19 pandemic precipitated the hunger situation globally and is said to expose additional 130 million people to the risk of acute hunger. In such a scenario, an intervention that could sustain the food security of the world and the nutritional needs of our population will be a boon.

The book "Achieving Zero Hunger" is an effort in this direction. It contains different chapters, and each chapter in itself is a comprehensive booklet. The chapters are lucid, *vi*vid and encompassing various strategies, policies, means and methods to increase global food production, its nutritious value, availability and accessibility in a sustained manner to achieve the zero hunger target. The contributions made by various international and national experts of repute have made this book a policy document in achieving zero hunger. With a strong basis in science and policy, this book will serve as a guiding torch and a tool kit for researchers, academicians, policymakers and students to pursue the global agenda of zero hunger.

**Prof. Nazir Ahmad Ganai**

*Vice Chancellor*
*SKUAST-K*

**SKUAST - KASHMIR**
National Agricultural Higher Education Project
Shalimar Srinagar 190025 | idpskuast@gmail.com, 0194-2461394

**STAR COLLEGE** "Excellence in Education"
Faculty of Veterinary Sciences & Animal Husbandry,
Alusteng, Srinager – 190006, J&K

# Preface

It was bizarre to see life coming to a standstill due to the covid-19 pandemic. Never before in my life had I observed such chaos and confusion among the masses, losing their kin, health and employment, and confining themselves in isolation. Worst hit were the vendors who used to earn their day-to-day living while reaching out to their buyers in every nook and corner. Food, the most essential commodity for survival was scarce, unavailable and out of reach of the needy. These odd scenes drove me to look into the hunger scenario in India. India, at the 102nd position in the Global Hunger Index, has a staggering 14 percent of the population going to bed hungry, 4.54 million babies born underweight and a 0.57 million of them dying every year because of under-nutrition. The pandemic is expected to lay many more millions hungry. The UN had chalked out a blueprint to end poverty and hunger, and to protect the planet to enjoy a peaceful and prosperous future through the Sustainable Development Goals (SDGs), with all United Nations Member States adopting 17 global goals in 2015. The global challenges we face include hunger, poverty, inequality, climate change, environmental degradation, peace and justice. SDG-2 targets to end hunger and ensure access, by people of all origins, to safe, nutritious and sufficient food all year round, to end all forms of malnutrition, double the agricultural productivity and incomes of farmers – ensuring sustainable food production systems. It also aims to implement resilient agricultural practices, strengthen capacity for adaptation to climate change, increase investment in rural infrastructure, agricultural research, extension services and technology development, and to prevent trade restrictions and distortions in world agricultural markets.

More than 60 percent of the world's population depends on agriculture for survival. With 50% of the habitable land being used for agriculture, its development is critically important to improve food and nutrition security. Since the food and agriculture sector offers key solutions for development and is pivotal to hunger and poverty eradication, I conceptualized an idea to sensitise the concerned people of the science, ways and means of ameliorating hunger in the country. Realizing the very essence of sustainable food production, accessibility of the population to it, rational consumption by consumers, trade and quality regulations, and the role of planning and policing in achieving zero hunger, I gathered experts from across the globe on one platform to exchange their views, thoughts and experiences, with an aim to provide some vital leads to rid our country free of hunger. These experts come from diverse backgrounds and specialize in different fields like agriculture, livestock, marketing, economics, trade, gender equity and government. With the grace of Almighty Allah, and the kind favour of reputed experts from across the globe, I could manage a 10-day workshop in favour of scientists from agriculture and allied fields from all over the country. In doing so, I was encouraged and financially supported by the Principal Investigator of the National Agriculture Higher Education Project (NAHEP- SKAUST-Kashmir), Prof. Nazir A. Ganai to organize the workshop and work towards a book derived from the lectures of these experts. As such, this book has been edited, compiled and carved out of the lectures delivered by reputed invited speakers at the workshop. It contains 22 chapters on different topics with the sole aim of achieving zero hunger in the country. The chapters in the book cover science, technology, skills and solutions related to almost all the targets of SDG-2. I am confident and of firm belief that this book will serve as a guide for agriculture scientists and academicians in building their capacity, skills and technologies, besides being a manual for planners and government functionaries to assist them in planning and designing the policies related to trade, marketing, quality and safety of food and food products. This book will hopefully carry out the resolve to pursue the agenda of zero hunger in the country. In this noble mission I thankfully acknowledge the support rendered by Dr. Aijaz A. Dar, Dr. Fozia Shah and Dr. Hena Hamdani. I would like to acknowledge and appreciate Astra Publications for their support and shouldering the noble responsibility of disseminating the knowledge/ science of achieving zero hunger all over the globe.

***Zahoor A. Pampori***

*Professor & Head*
*Division of Animal Physiology, FVSc&AH,*
*SKUAST-K, Shuhama, Alusteng, Srinagar, Kashmir*

# Contents


*Acknowledgment*     *v*

*Foreword*     *vii*

*Preface*     *ix*

1. **Future Smart Food: Rediscovering Hidden Treasures of Neglected and Underutilized Species for Zero Hunger in Asia**     **1**
*Kadambot H.M. Siddique*

2. **Nutrition Sensitive Agriculture (NSA) and Food Systems**     **15**
*Vinay Singh*

3. **Creating Sustainable Value Chains for Food System Transformation in South Asia**     **29**
*Nafees Meah*

4. **Sustainable Agriculture Practices for Food Security and Zero Hunger in India**     **37**
*Ch. Srinivasa Rao*

5. **The Achievement of Food and Nutrition Security in South Asia is Deeply Gendered**     **43**
*Nitya Roa*

6. **Focus on Sustainable Agricultural Intensification and Small-scale Agriculture for Achieving Zero Hunger**     **53**
*P.V. Vara Prasad*

*xii*

7. **The Bumpy Journey of Food and Nutrition Security in India**    67
   *A.K. Srivastava*

8. **Production, Food Supply Chains, Trade and Nutrition: The Case of India**    77
   *Arpita Mukherjee, Angana Parashar Sarma and Nibha Bharti*

9. **Overview of Child Health and Nutrition Programme in India**    95
   *Sila Deb*

10. **Importance of Agricultural Trade and Regional Cooperation for Food Security**    103
    *Joseph George*

11. **Sustainable Agricultural Research and Technology: Needs to Address Food Security**    109
    *Zahoor A. Ganie*

12. **Crop Biofortification for Achieving Zero Hunger by 2030**    113
    *Z.A. Dar, B. Kumar, R. Munshi, S. Naseer, F. Rasool, A.A. Lone, N.S. Khuroo, S.A. Dar, I. Abidi, G. Ali, A.M. Iqbal and A. Khan*

13. **Stress Resilient Crop Production for Sustainable Agriculture**    127
    *R.H. Kanth, O.A. Wani and S. Fayaz*

14. **Redefining Food Security: Why Innovation is Not Enough**    139
    *Puja Theil*

15. **How Countries Can Reduce Child Stunting at Scale: Lessons from Exemplar Countries**    143
    *Zulfiqar A. Bhutta*

16. **Role of Livestock in Alleviating the Burden of Hunger by 2030**    165
    *B.S. Prakash*

17. **Animal Health, a Prerequisite for Better Human Health: One Health Perspective**    183
    *S.V.S. Malik*

18. **Dairy Food Production Interventions for Livelihood Support and Food and Nutritional Security**    197
    *Mohammad Ashraf Paul*

19. **Aquaculture: Mitigating Global Hunger**    205
    *M.H. Balkhi, Anayit ullah Chesti and Mansoor Rather*

**20. Achieving Zero Hunger Target: Innovative Solutions through Science and Technology**  211

*Sheikh Firdous Ahmad*

**21. Strategies to Achieve Zero Hunger through Agricultural Technologies**  219

*Vijaymahantesh and Vijaya R. Chitnis*

**22. Safety and Quality Assurance of Animal Food Products: Key to Achieving Zero Hunger**  223

*Atul Kumar*

**23. Battle of Hunger in India: Efforts, Limitations, Possessions and the Way-Forward**  235

*Zahoor A. Pampori*

**Glossary**  261

*Index*  267

# Future Smart Food: Rediscovering Hidden Treasures of Neglected and Underutilized Species for Zero Hunger in Asia

**Kadambot H.M. Siddique**

*Director,*
*The UWA Institute of Agriculture, Australia*

## Abstract

*In a world that talks of "obesity epidemics" and "global connectedness", hundreds of millions struggle each day to get enough food to survive. For more than 800 million people on our planet, much of their daily life and what energy they have is spent in an exhausting, debilitating search for food. A hundred million children are underweight. They are losing the battle. More than three million of them will starve to death this year.*

The UN Sustainable Development Goal-SDG2 (Zero Hunger goal) has specific targets for doubling agricultural productivity and ensuring sustainable food systems. We know that to do this, we have to improve the incomes of small-scale food farmers and we have to maintain genetic diversity.

There are almost 30,000 edible plant species, but 90 per cent of the calories in the human diet comes from just over a hundred of them. These 30,000 species include many we refer to as NUS: neglected or underutilised species. Some are wild;

*(Part of it published as Executive Summary from Food and Agriculture Organization of the United Nations, Bangkok, 2018 by Kadambot H.M. Siddique and Xuan Li)*

some are semi-domesticated; all are under-utilised. As a rule, they do not attract the attention of commodities markets. Working with local people, we are hoping to change that. We are re-imagining some of these NUS crops as 'Future Smart Foods'. To make the grade, Future Smart Food crops have to be nutritionally dense, climate resilient, economically viable, and locally available and adaptable. If we are going to feed our 800 million undernourished people, we must close two significant gaps.

1. The first is a production gap: Agricultural production will have to increase globally by at least 50 per cent to meet food demand by 2050.

2. The second is a nutrition gap: The gap between the crops grown for commercial global market and the foods that are needed for a healthy diet.

We are looking at crops, which can be grown at the local and regional level, where they contribute to environmental sustainability, as well as increasing agricultural production and improving nutritional outcomes. Some studies have predicted that climate change in South Asia will result in declines of 14 per cent in rice production, 9-19 per cent in maize production and 44-49 per cent in wheat production. We have to make a move on Future Smart Food crops. Traditional staple crops (such as rice, wheat and maize) are not only at threat from climate change; they do not provide enough of the wide range of nutrients required for a healthy diet.

Scoping and prioritisation studies on these Future Smart Foods are currently being conducted in Bangladesh, Bhutan, Cambodia, Lao People's Democratic Republic (PDR), Myanmar, Nepal, Vietnam, and West Bengal. They are working with crops we know reasonably well: like beans, lentils, peas, peanuts, pumpkin, quinoa, buckwheat, soybeans, sweet potatoes, taro, walnuts and yams. Nevertheless, there are all sorts of yams, like the purple and fancy yams of the Lao PDR, the greater yam of Vietnam, and the elephant foot yam of Myanmar and West Bengal. In addition, there are mungo and rice beans, grass peas and cowpeas. Then there are the crops we may never have tried, although we have heard of them: crops like buckwheat, quinoa and sorghum. We are also working with drumstick — but the leaves, flowers and immature pods of the moringa tree (very high in vitamin C), in Bhutan, Cambodia, Myanmar, Nepal and West Bengal.

In Bangladesh, we are working with crops like foxtail millet, an ancient food source, and snake gourd. Cambodia has ivy or scarlet gourds, cooked as a vegetable.

Myanmar is exploring the possibilities of roselle, which is also in folk medicine. Nepal has jackfruit, which can weigh as much as 55 kilos each, and the butter tree, which produces ghee for cooking, fuel, manure and body lotion, as well as providing an interesting form of alcohol.

In West Bengal, they are doing things with style: aromatic rice and swamp taro, as well as drumstick and jackfruit. They are also working with black and green gram, amaranthus (rich in Vitamins A and C, calcium and manganese), and fenugreek.

Hunger, food insecurity and malnutrition are major challenges in the twenty-first century for Asia and Pacific. To achieve Zero Hunger, which stands at the core of the SDGs, food systems need to be improved urgently, and dietary patterns have to change. There are disconnects in the value chain between production, consumption and nutrition. Current agricultural production patterns are not sustainable, are unlikely to achieve necessary growth rates, and do not offer the right mix of nutrients needed to healthy diet. Addressing hunger and malnutrition in a changing climate is considered a top priority by most countries in Asia. Within an agricultural diversification and sustainable intensification strategy, promoting Future Smart Food through a food systems approach is a cost-effective intervention to address the dual challenge of malnutrition and climate changes.

United Nation's Sustainable Development Goal 2 (SDG 2) - Zero Hunger calls for the eradication of hunger and all forms of malnutrition, with targets for doubling agricultural productivity and incomes of small-scale food producers (SDG 2.3), ensuring sustainable food systems (SDG 2.4) and maintaining genetic diversity (SDG 2.5). Promoting NUS could be a powerful means of achieving the Zero Hunger goal while offering solutions to some worrying trends in agriculture.

From a demand perspective, the multiple challenges for global agriculture and food systems include population growth, increasing urbanization and the emergence of a larger middle-class, which has given rise to new food preferences and changing consumer attitudes that involve concerns about food quality and safety. Taken together, these trends will have a huge impact on Asia's future dietary patterns. From a supply perspective, there is concern about the slowing down of yield growth in staple crops to levels that are insufficient for meeting future food demands without the expansion of agricultural land, which is already scarce in Asia. The combined effects of climate change, declining agricultural biodiversity, water scarcity and degradation of natural resources are challenging world food security. Some studies predict that in South Asia, the climate change scenario would result in a 14 per cent decline in rice production relative to the no-climate-change scenario, a 44-49 per cent decline in wheat production, and a 9-19 per cent fall in maize production (Nelson *et al.*, 2009). There are two significant gaps that exist, or will emerge, in our agriculture and food systems:

1. **Production gap:** A 30 per cent increase in the global population by 2050 will require a 60-70 per cent increase in food production, taking into account changing consumption patterns. Production increases of traditional staple crops are unlikely to meet the increasing demand; irrigated wheat, rice and maize systems appear to be near 80 per cent of the yield potential. Therefore, relying on these crops alone will not be enough to close the gap between food supply and demand.

2. **Nutrition gap:** Even if traditional staple crops provide enough calories to prevent hunger, they do not provide all the nutrients necessary for a healthy diet. Current high levels of malnutrition are often due to unbalanced diets with insufficient nutrition diversity. Closing the production and nutrition gaps requires a transformation of current agriculture and food systems towards greater diversity.

A holistic food system perspective can provide answers for tackling malnutrition, and addressing climate change and environmental threats in agricultural production.

Nutrition-sensitive and climate-smart agriculture interventions can tap local potential to promote agricultural productivity that meets nutritional requirements. These interventions will go beyond the promotion of current staple crops and include crops previously considered of secondary importance.

Characteristically, neglected and underutilized species (NUS) are nutritious, climate resilient, economically viable (in the right setting) and adapt to local conditions, especially in marginal areas. In the past, NUS have been ignored by agricultural research, not included in agricultural extension curricula, and did not benefit from organized value chains. However, due to their adaptability and nutritional qualities, many NUS could make a major contribution to increased food availability, affordability and nutrition security. Recognizing this potential, FAO RAP's Regional Initiative on Zero Hunger Challenge (RI-ZHC) has embarked on the promotion of NUS crops as a means to foster food and nutrition security, although the definition of NUS also includes livestock, fisheries and aquaculture species.

## Agriculture and Food Systems

There is a clear connection between the over-reliance on a few staple crops, low dietary diversity and malnutrition in agricultural and food systems at the country level. A leading cause of persistent malnutrition is poor dietary diversity (poor quality and limited variety of food in the diet). Dietary diversity is low when a high consumption of cereals is accompanied by a low intake of vegetables, fruits and pulses, which could provide the necessary micronutrients and fibre. This dependency leads to a significant nutrition gap. Agricultural production in Asia focuses on a few staple crops, particularly rice. The pattern reflects a structural issue: too many people consume food with too few nutrients and too much food is being produced without offering enough nutrients. This is often the involuntary consequence of government policies that prioritize quantitative food production targets.

Figure 1.1 highlights that agricultural diversification is a powerful tool for achieving Zero Hunger. It outlines why we need diversity on two counts: dietary diversity, which is a cost-effective, affordable and sustainable way to prevent hunger and malnutrition; and product diversity, which makes it possible to supply nutritious and diversified food that provides better options for dealing with changing environments, especially the effects of climate change.

There are two major limiting factors in global agriculture and food systems, both of which are observed in Asia:

1. Limited production diversity with an emphasis on starchy crops can lead to unbalanced diets and ultimately malnutrition. An abundant supply of a few staple crops alone does not provide sufficient nutrition.

2. Reliance on a few staple crops with high input requirements leaves farming more vulnerable to environmental shocks, especially under a climate change scenario.

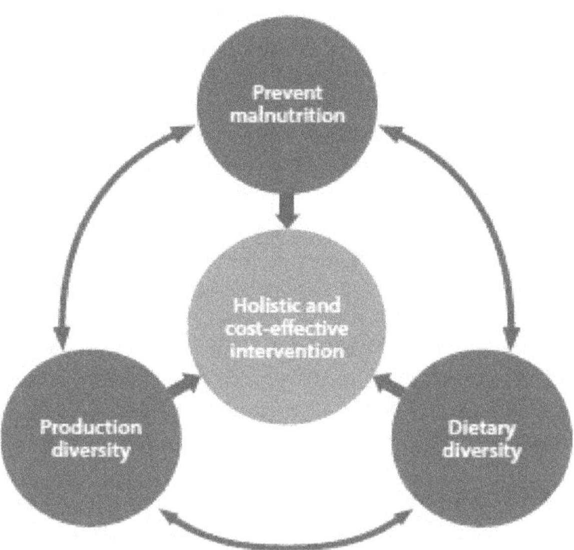


**Figure 1.1 Highlights that Agricultural Diversification is a Powerful Tool for Achieving Zero Hunger.**

According to Graziano da Silva (FAO, 2012a), dependence on a few crops has negative consequences for ecosystems, food diversity and health. Food monotony increases the risk of micronutrient deficiency.

## Neglected and Underutilized Species as Entry Points for Addressing Malnutrition from a Food System Perspective

From a food system perspective, dietary and production diversity need to improve to address malnutrition. NUS are underexplored and can be called 'hidden treasures' that offer tremendous opportunities for fighting poverty, hunger and malnutrition. As the FAO Director-General highlighted, NUS play a crucial role in the fight against hunger and are a key resource for agriculture and rural development (FAO, 2012a). Considering the wide coverage of NUS including crops, livestock, fisheries and aquaculture, and forests, this regional priority-setting exercise set crops as an entry point among NUS to address hunger and malnutrition. Historically, underutilized plants have been used for food and other uses on a large scale and, in some countries, are still common especially among small or marginal farmers in rural areas where many are traded locally, and a lucky few NUS have made their way to export niche markets around the world (Akinnifesi *et al.*, 2008). NUS have high nutritional value and can be an essential source of micronutrients, protein, energy and fibre, which contribute to food and nutrition security. Apart from their superior nutritional qualities, many of these crops do not require high inputs, can be grown on marginal lands and easily intercropped or rotated with staple crops, as well as fit easily into integrated practices such as agro-ecology. Because they are frequently adapted to marginal conditions, and many have the unique ability to tolerate or withstand stresses, NUS can make production systems more sustainable and climate resilient.

## Global Policy Frameworks and the Integration of Neglected and Underutilized Species

The importance of NUS is widely recognized by the global scientific community (Joint FAO/IAEA 2004; Kahane *et al.*, 2013; Khoury *et al.*, 2014; Nyadanu *et al.*, 2016; Rutto *et al.*, 2016; Stamp *et al.*, 2012). The development and implementation of policies are often a key component in promoting NUS/FSF into agricultural production systems (Noorani *et al.*, 2015). This section provides an overview of the policy frameworks. It includes the first Global Plan of Action (GPA) for the Conservation and Sustainable Use of Plant Genetic Resources for Food and Agriculture (PGRFA) (FAO, 1996) adopted by 150 countries in 1996; the International Treaty on Plant Genetic Resources for Food and Agriculture (the Treaty) that entered into force in 2004, which provides a legal framework whereby governments, farmers, research institutes and agro-industries can share and exchange PGRFA and benefits derived from their use (FAO, 2009); the Second Global Plan of Action for Plant Genetic Resources for Food and Agriculture (Second GPA) in 2011 (FAO, 2012b); the Cordoba Declaration (FAO, 2012c), which was elaborated upon at the international seminar on Crops for the XXI Century, and further emphasized the importance of underutilized and promising crops for the international arena; and the Second International Conference on Nutrition (ICN2) held in Rome in 2014, which showcased the profile of NUS and adopted the Rome Declaration on Nutrition. This high-level conference emphasized the importance of NUS through Recommendation 10: *"Promote the diversification of crops including underutilized traditional crops, more production of fruits and vegetables, and appropriate production of animal-source products as needed, applying sustainable food production and natural resource management practices."* All these policies reflect recent trends and calls for a global commitment to enhance the conservation and sustainable use of NUS/FSF, and for these frameworks to be translated into actions (see Figure 1.2 for a timeline of relevant international policy developments).

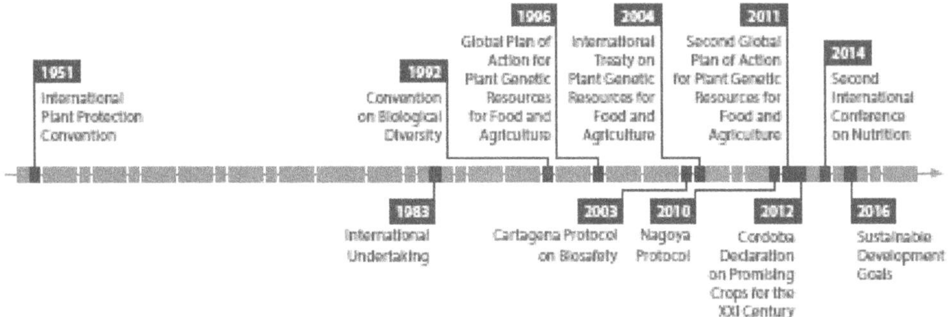


**Figure 1.2: Timeline Showing the Major Relevant International Policy Frameworks for the Conservation and Sustainable Use of Plant Diversity.**

# Regional Priority Setting Exercise on NUS: Future Smart Food

Given the wide range of NUS and their diverse characteristics and potential benefits, the Zero Hunger Initiative started by scoping, identifying and prioritizing promising NUS at the country level to create a list of NUS capable of helping to close production and nutrition gaps. Apart from their advantages for nutrition and production, the selected NUS also needed to be economically viable and socially acceptable. The criteria for prioritizing NUS were established in four categories:

1. Nutrition (nutritional value and health benefits);
2. Production (local knowledge, availability, seasonality, productivity, intercropping and competition from other crops, and processing);
3. Ecology (agro-ecology, adaptability to potentially changing local climates and soil types), and
4. Socio-economy (cultural acceptance and consumer preferences, access to markets and potential income generation).

Each participating country conducted assessments according to these criteria. The regional priority-setting exercise focused on the following groups:

- ☆ Cereals,
- ☆ Horticultural species,
- ☆ Nuts and pulses,
- ☆ Roots and tubers, and
- ☆ Others.

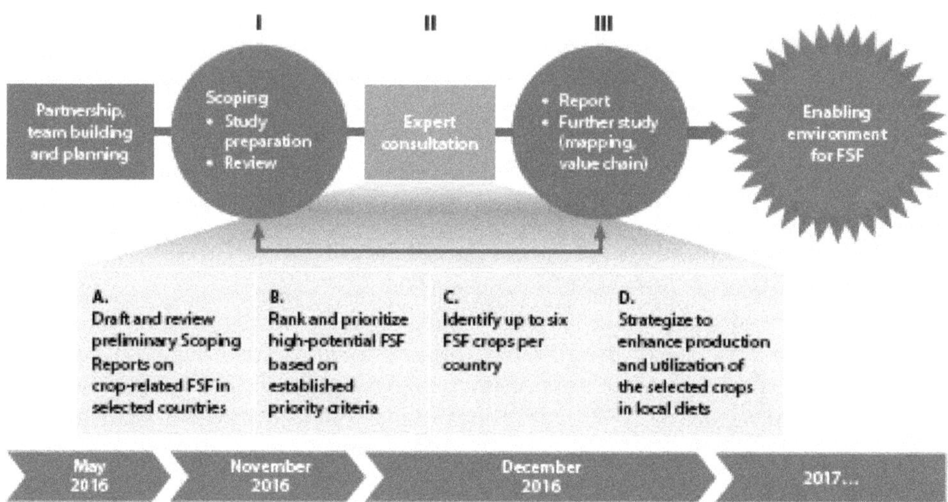


**Figure 1.3: Regional Priority-Setting Exercise on Scoping, Prioritizing and Mapping of NUS.**

Scoping of NUS was limited to the available species in the national gene bank. Prioritization followed the principles of country ownership and country specificity. As such, the NUS scoping and prioritization results are owned by the participating countries. International partners offered technical assistance through multidisciplinary reviews and verifications to support each country. The status of NUS is country specific, which means that a species recognized as NUS in one country may not be considered as NUS in another country. Following these principles and criteria, FAO – in collaboration with FAO's Special Ambassador on International Year of Pulses, the University of Western Australia, ICARDA, ICRISAT, MSSRF-LANSA, CATAS-TCGRI, Mahidol University, ACIAR, ICIMOD, CFF, as well as national governments and research institutes – conducted an interdisciplinary priority-setting exercise (Figure 1.3).

Based on the national scoping studies and international interdisciplinary review, each country prioritized up to six promising NUS as candidates for FSF. The initial results of the FSF selection process in eight countries in South and Southeast Asia (Bangladesh, Bhutan, Cambodia, Lao PDR, Myanmar, Nepal, Viet Nam and West Bengal in India) are presented in Table 1.1.

**Table 1.1: Potential Future Smart Food in Eight Countries in South and Southeast Asia**

| Cereals | Roots and tubers | Pulses | Fruits and vegetables | Nuts, seeds and spices |
|---|---|---|---|---|
| Amaranth | Elephant foot yam | Black gram | Chayote | Linseed |
| Buckwheat | Fancy yam | Cow pea | Drumstick | Nepali butter tree |
| Finger millet | Purple yam | Faba bean | Fenugreek | Nepali pepper |
| Foxtail millet | Swamp taro | Grass pea | Indian gooseberry | Perilla |
| Grain amaranth | Sweet potato | Horse gram | Jackfruit | Walnut |
| Proso millet | Taro | Lentil | Pumpkin | |
| Quinoa | | Mung bean | Roselle | |
| Sorghum | | Rice bean | Snake gourd | |
| Specialty rice | | Soybean | Wood apple | |
| Tartary buckwheat | | | | |

# Challenges, Opportunities and Strategies for Neglected and Underutilized Species as Future Smart Food for Zero Hunger

Mahmoud Solh argues that the projected 50 per cent increase in demand for agricultural production from 2013 to 2050 (FAO, 2017) has to be achieved despite the ongoing degradation of natural resources and the serious implications of climate change on agriculture. Despite impressive progress, more than 800 000 people are still facing hunger globally (FAO, 2017), and many Asian countries are categorized as 'serious' in the Global Hunger Index (GHI) because 20 per cent to 34 per cent of their populations are undernourished. NUS have important advantages over widely grown commercial crops. They often link with the cultural heritage of their places of origin; are well adapted to specific agro-ecological niches, harsh environments and marginal lands; and perform well in traditional production systems with little or no external inputs. However, NUS have been given little attention by national research, extension services, policy and decision makers, donors, technology providers and consumers. NUS crops are scarcely represented in *ex-situ* collections and, despite having nutritious and/or medicinal properties or other multiple uses, are often not attractive to the public and private sectors due to social, economic, environmental (*e.g.* genetic erosion of NUS gene pools), agronomic (*e.g.* lack of seed supply systems) or political reasons.

There are several examples of NUS that have become important commercial crops. Quinoa and lentil are two NUS crops that attracted global interest despite being important mainly in subsistence agriculture locally or regionally. Quinoa used to be a subsistence crop in the Lake Titicaca basin of Peru and Bolivia, but with rising consumer appreciation, its production in Bolivia, Peru and Ecuador almost tripled from 1992 to 2010. Quinoa cultivation has spread to more than 70 countries, including Denmark, England, France, Holland, India, Italy, Kenya, Sweden and even Australia. The reasons for the fast evolution of quinoa as a commercially important food are its high nutritional value, good taste and quick cooking properties. Quinoa was promoted globally with 2013 being the International Year of Quinoa; but to reach this stage, quinoa underwent numerous phases:

☆ Perception change among farmers, processors, policymakers and the public in general

☆ Improved capacity of researchers, educators, policymakers, farming communities and the private sector

☆ Better documentation of indigenous knowledge and use, propagation, growth characteristics, nutritional aspects, adaptation to harsh environment, resistant traits and variability

☆ Improved *ex-situ* and *in-situ* conservation

☆ Breeding programmes to enhance productivity and quality traits

☆ Development of seed production systems

☆ Involvement of stakeholders at all stages in the value chain

☆ Development of the value chain and exploration of markets at local, national and international levels

☆ Modernization of agricultural production including mechanization

☆ Development of post-harvest technology and conservation

☆ Creation of supportive policy environments including legal frameworks at both national and international levels (incentives such as subsidies to promote targeted species)

☆ Increased cooperation at national, regional and international levels

Lentil, along with other pulses, is known as 'poor man's meat' in the Middle East, North Africa and South Asia due to its relatively high protein content. In addition to this important nutritional advantage, lentil is well adapted to low input agriculture and its ability to fix atmospheric nitrogen enriches soil fertility and ensures agricultural sustainability.

In Ethiopia, India and Turkey, lentils and other pulses are becoming cash crops, as demand has been rising for local production due to government interventions and investments in research and technology transfer with national and international partners. International agricultural research centres, such as ICARDA, in collaboration with national agricultural research systems, worked, in particular, on better crop management and improved varieties, improved crop genetics, horizontal expansion of lentil production in rice–fallow replacement, reduction of post-harvest losses, and improved plant architecture and mechanical harvesting.

Particularly in South Asia, lentil development has bridged yield gaps by 65 per cent to 75 per cent in farmers' fields as a result of improved production packages in rainfed agriculture. Lentil cultivars with high concentrations of iron and zinc have been released in Bangladesh, Ethiopia, India, Nepal, Portugal, Syria and Turkey. In Bangladesh, five bio-fortified varieties covered 90 000 ha or 43 per cent of the total lentil cultivated area in 2014.

These examples show that NUS are important for conserving and enhancing biological diversity and protecting fragile ecosystems. However, the potential of NUS goes beyond local niches: they have become globally important both as a means to achieving greater food and nutrition security, and as a response to dealing with the implications of climate change.

Kadambot H.M. Siddique and Xuan Li highlights the potential of FSF for nutrition enhancement and climate resilience, and points out the compelling need to rethink agricultural strategies in a climate-change context. Production and nutrition gaps will widen if conventional production patterns continue. Finding solutions to bridge the production and nutrition gaps cannot be achieved without a fundamental transformation of the current agricultural and food system. Agricultural diversification with sustainable intensification is a promising strategy for achieving Zero Hunger in a climate-change context, where NUS have a significant role to play.

Most NUS do not require high inputs and can be grown on marginal and degraded lands while contributing to increased agricultural production, crop diversification and a better environment. In turn, food and nutrition security are improved. Pulses, in particular, are a potential health food, reducing risks of chronic diseases, obesity and other conditions. The current over-reliance on a handful of major staple crops (rice, wheat, maize and potato) has inherent agronomic, ecological, nutritional and economic risks. National and international research bodies, non-government organizations, community-based organizations and commercial entities interacting with farmers have not paid enough attention to NUS. However, not all NUS will effectively foster food and nutrition security. NUS have to be assessed and prioritized following established criteria. NUS will be labelled FSF if they are:

1. Nutritionally dense,
2. Climate resilient,
3. Economically viable, and
4. Locally available or adaptable.

Pulses are an excellent example of an FSF that exhibits all four criteria.

The chapter shows that the promotion of FSF will:

☆ contribute to closing nutrition gaps and offer food security and nutrition;

☆ reduce the risk of over-reliance on limited numbers of staple crops and increase the sustainability of agriculture by reducing inputs;

☆ provide focused effort to help marginalized and indigenous people improve their livelihoods and income; and

☆ contribute to the preservation and celebration of cultural diversity.

These characteristics amply justify investments in NUS, as they have enormous potential for bridging the dual production and nutrition gaps. In light of the multi-functionality of NUS, it is time to rediscover these 'hidden treasures'.

## Way Forward

Hunger, food insecurity and malnutrition are major challenges in the twenty-first century for Asia and the Pacific. To achieve Zero Hunger, which stands at the core of the SDGs, food systems need to be improved urgently, and dietary patterns have to change. There are disconnects in the value chain between production, consumption and nutrition. Current agricultural production patterns are not sustainable, are unlikely to achieve necessary growth rates, and do not offer the right mix of nutrients needed for a healthy diet. Consumers are often unaware of their potential food choices and stick to dietary patterns that do not agree with an urbanized lifestyle. Poor people's diets are still over-dependent on a few staple crops that lack important nutrients and minerals. At the same time, the newly affluent strata of society have begun to suffer from obesity and related health issues.

Agricultural diversification and resilience offer enormous opportunities for addressing hunger and malnutrition, especially in the context of climate change. In this regard, NUS offer diverse and nutritious food resources. NUS are important in specific agro-ecological niches, often linked with tradition and cultural heritage in their places of origin. They are an essential source of protein and micronutrients, they enhance climate resilience, improve agriculture sustainability, and boost household income and livelihoods with considerable commercial potential.

Regional Expert Consultation on Scoping, Prioritizing and Mapping of NUS, organized by FAO in collaboration with national and international partners in Bangkok from 3-5 December 2016 with panel of experts, representing eight countries, as well as 22 national and international partners developed ten key recommendations:

1. Urgent call for decision makers to raise awareness of the nutrition-sensitive and climate-resilient benefits of NUS to address hunger, malnutrition and climate change.

2. Recognize, identify and promote complementarities of NUS with existing staple crops for nutrition enhancement, climate-change resilience and diversification of cropping systems, and reliable NUS as Future Smart Food (FSF) to popularize these species.

3. Establish a National Coordinating Committee on FSF involving concerned ministries and appoint a Strategic Coordinator at the inter-ministerial level.

4. Create an enabling environment by strengthening national institutional support for mainstreaming FSF into national policies and programmes, using appropriate incentives, procurement of FSF for food programmes (*e.g.* mid-day meal/school-meal scheme) to enhance national consumption, local production and facilitate marketing.

5. Establish nationally coordinated research for development programmes targeting FSF with high potential, and expand coverage of national agriculture statistics and national food composition data on FSF for evidence-based decision making.

6. Document and validate best-bet FSF case studies, compile indigenous knowledge related to FSF, undertake clinical and field studies to demonstrate the health benefits and climate resilience of FSF, and assemble quantitative data for public dissemination.

7. Enhance public awareness of the importance of FSF by developing nutrition and climate-change education materials and curricula on the importance of FSF for consumers, traders, producers, health professionals, researchers, teachers (*e.g.* school curricula), farmers, women and youth.

8. Identify key entry points in the value chain and encourage value-chain development for specific FSF, including innovative and targeted interventions for promotion (*e.g.* ready-to-use food products) and increase funds for research, development and extension capacities on FSF production and processing technologies.

9. Strengthen multidisciplinary and multi-sectoral collaboration through existing coordination mechanisms and build partnerships at national and regional levels, including academia, civil society and the private sector, to enhance research and consumption and to attract the private sector to boost production, processing, value addition, product development and marketing of FSF.

10. Establish a regionally coordinated network on FSF to facilitate the exchange of information, policy, technologies and genetic resources, as well as FSF promotion, in target countries.

Addressing hunger and malnutrition in a changing climate is considered a top priority by most countries in Asia. Within an agricultural diversification and sustainable intensification strategy, promoting FSF through a food systems approach is a cost-effective intervention to address the dual challenge of malnutrition and climate change. An enabling environment for promoting FSF is essential. This requires a long-term vision and holistic approach that integrate political, economic and environmental aspects, multi-stakeholder cooperation, and forward-looking institutions covering production, marketing and consumption stages.

## REFERENCES

Akinnifesi F.K., Leakey, R.R.B., Ajayi, O.C., Sileshi, G., Tchoundjeu, Z.,

Matakala, P. and Kwesiga, F.R. 2008. *Indigenous Fruit Trees in the Tropics: Domestication, Utilization and Commercialization*. CAB International Publishing, Wallingford, UK, p. 438.

FAO. 1996. Global Plan of Action for the Conservation and Sustainable Utilization of World's Plant Genetic Resources for Food and Agriculture and Leipzig Declaration, adopted by the International Technical Conference on Plant Genetic Resources, Leipzig, Germany, 17-23 June 1996. FAO, Rome.

FAO. 2009. International Treaty on Plant Genetic Resources for Food and Agriculture. FAO, Rome. (also available at www.planttreaty.org/content/texts-treaty-official-versions).

FAO. 2012a. Neglected crops need a rethink – can help world face the food security challenges of the future. (also available at www.fao.org/news/story/en/item/166368/icode/).

FAO. 2012b. The Second Global Plan of Action for Plant Genetic Resources for Food and Agriculture. FAO, Rome. (also available online at http: //www.fao.org/docrep/015/i2624e/i2624e00.htm).

FAO. 2012c. Cordoba Declaration on Promising Crops for the XXI Century. (also available at www.fao.org/fileadmin/templates/food_ composition/documents/Cordoba_NUS_Declaration_2012_ FINAL.pdf).

FAO. 2017. The Future of Food and Agriculture, Trends and Challenges. Rome, 2017.

FAO and IAEA. 2004. Division of Nuclear Techniques in Food and Agriculture, Vienna. Genetic improvement of under-utilized and neglected crops in low income food deficit countries through irradiation and related techniques Proceedings of a final research coordination meeting (IAEA-TECDOC-1426). International Atomic Energy Agency.

Kahane R., Hodgkin T, Jaenicke, H, Hoogendoorn C, Hermann M, Hughes JDA, Padulosi S, and Looney N. Agrobiodiversity for food security, health and income. *Agronomy for sustainable development* 33(4), pp. 671-693.

Khoury, C.K., Bjorkman, A.D., Dempewolf, H., Ramirez-Villegas, J., Guarino, L., Jarvis, A. Rieseberg, L.H. and Struik, P.C., 2014. Increasing homogeneity in global food supplies and the implications for food security. *Proceedings of the National Academy of Sciences*, 111(11), pp. 4001-4006.

Nelson, G. C., Rosegrant, M. W., Koo, J., Robertson, R., Sulser, T., Zhu, T. and Magalhaes, M. 2009. Climate Change: Impact on Agriculture and Costs of Adaptation. *Food Policy Report. International Food Policy Research Institute (IFPRI), Washington, D.C.*, pp. DOI: 10.2499/0896295354.

Noorani A., Bazile D., Diulgheroff S., Kahane R. and Nono-Womdim R. 2015. Promoting neglected and underutilized species through policies and legal frameworks. De Ron, Antonio M. ed. EUCARPIA *International Symposium on Protein Crops: V Meeting AEL. Pontevedra.* pp. 107-111.

Nyadanu, D., Aboagye, L.M., Akromah, R. and Dansi, A., 2016. Agro-biodiversity and challenges of on-farm conservation: the case of plant genetic resources of neglected and underutilized crop species in Ghana. *Genetic Resources and Crop Evolution*, 63(8), pp.1397-1409.

Rutto, L.K., Temu, V.W. and Ansari, M.S. 2016. Genetic Vulnerability and Crop Loss: The Case for Research on Underutilized and Alternative Crops. *Mathematical Sciences with Multidisciplinary Applications*. pp. 465-479. Springer International Publishing.

Stamp, P., Messmer, R. and Walter, A. 2012. Competitive Underutilized crops will depend on the state funding of breeding programmes: an opinion on the example of Europe. *Plant breeding*, 131(4), pp.461-464.

## Chapter 2

# Nutrition Sensitive Agriculture (NSA) and Food Systems

Vinay Singh

*Food Security and Nutrition Expert,*
*FAO of the United Nations, New Delhi*

## Nutrition Sensitive Agriculture (NSA)

Nutrition sensitive agriculture is an approach that seeks to ensure the production of a variety of affordable, nutritious, culturally appropriate and safe foods in adequate quantity and quality to meet the dietary requirements of populations in a sustainable manner. The recognition that addressing nutrition requires taking action at all stages of the food chain - from production, processing, retail to consumption – has led to a broader focus which encompasses the entire food system.

## Importance of NSA

Nutrition-sensitive agriculture stresses the multiple benefits derived from enjoying a variety of foods, recognizing the nutritional value of food for good nutrition, and the importance and social significance of the food and agricultural sector for supporting rural livelihoods. The overall objective of nutrition-sensitive agriculture is to make the food system better equipped to produce good nutritional outcomes.

## Agriculture and Food System

In fact, food systems encompass all the people, institutions and processes by which agricultural products are produced, processed, and brought to consumers (FAO SOFA 2013). "A food system gathers all the elements (environment, people, inputs, processes, infrastructures, institutions, *etc.*) and activities that relate to the production, processing, distribution, preparation and consumption of food, and the outputs of these activities, including socioeconomic and environmental outcomes." (HLPE 2014, p29)

Food system functions determine availability, affordability, convenience and desirability of various foods – in other words, the food environment – and thus the behaviour of consumers. Through the food environment, the food system influences consumers' diets and nutritional status.

## Interventions to Make Agriculture and Food Systems Nutrition-Sensitive

In order to improve nutrition and to effectively address the causes of malnutrition, working across the food system functions is very important. The entry point for interventions to make agriculture and food systems nutrition-sensitive are organized according to four key functions of the food system and as cross-cutting issues. They are as follows:

### 1.  Food Production

### Diversification and Sustainable Intensification of Agricultural Production

The precondition for good nutrition is that a diversity of foods is available and affordable for all individuals at all times. However, the global food system is currently not meeting global requirements for the production of adequate amounts of nutritious foods necessary for healthy diets. At local level, excessive intensification (*i.e.* monoculture) risks simplifying diets and worsening nutrition in producer communities and threatens ecosystem resilience.

Diversification on a large scale (*e.g.* implemented at regional or national level and/or involving commercially oriented producers) can help enhance availability of diverse foods in markets and reduce prices of nutritious foods. Integrated farming systems (*e.g.* legume-based cropping systems including crop rotation and intercropping, rice-wheat farming systems) favour both diversification and sustainable intensification of production. Home gardening with emphasis on nutrient-dense varieties of vegetables and fruit trees and small-scale integrated farming systems (*e.g.* mixed crop-livestock-aquaculture systems or VAC systems) have demonstrated potential to improve diet quality and raise levels of nutrition for producing households.

### Nutrition-sensitive Livestock and Fisheries

The inclusion of animal source foods (ASFs) in the diet is an important food-based strategy for improving and safeguarding nutrition. In addition to being rich in protein and energy, animal source foods can be an excellent source of selected micronutrients (easily absorbable iron, zinc, calcium, vitamin A, vitamin B12 and various essential amino-acids). Fish products can also be good and natural sources of long chain omega-3 and iodine, both important for optimal brain development in children.

Livestock ownership (*e.g.* cattle, chicken and other poultry, small ruminants such as goats and sheeps) can contribute to dietary diversity and nutritional outcomes through home consumption and income generation, especially if accompanied with nutrition education aimed at promoting consumption of ASFs including for complementary feeding. Milk and dairy production is often used as a strategy to enhance income and livelihoods. Nutrition objectives can easily be integrated, for example by ensuring consumption by children or linking with school meal programs.

## Biodiversity for Food and Nutrition

Biodiversity can play a key role in ensuring dietary diversity and assuring nutrient adequacy. For example, some varieties of bananas can contain up to 1000 times more pro-vitamin A carotenoid then the most globally consumed variety. Therefore, the intake of one variety rather than another can mean the difference between micronutrient deficiency and micronutrient adequacy.

Selection and production of species and varieties should be based not only on yields but also on nutrient content (concept of nutrient productivity), thereby enhancing the nutrient supply of agricultural products, especially for micronutrients. Community-level initiatives for supporting the saving and exchange of seeds (*e.g.* community seed banks, village seed fairs, smallholder seed enterprises) and protecting ecosystems (*e.g.* community-based natural resource management, reforestation, promotion of micronutrient-rich forest foods) enhance availability of and access to genetic resources, strengthen local food systems and empower indigenous people.

## Biofortification

While it is important to continue efforts to increase dietary diversity and quality as a long-term solution to all forms of malnutrition, consumption of biofortified crops allows many people to increase dietary micronutrient adequacy simply by substituting a micronutrient-poor staple with its micronutrient-rich counterpart. The main target group of biofortification programmes is subsistence and semi-subsistence farmers who grow crops for their own consumption. To achieve impact beyond the farm gate, biofortification programmes typically include additional activities such as technical assistance in post-harvest storage and handling, creation of market linkages, and support to value addition and demand creation. These post-farm gate activities require building a strong network of stakeholders all along the value chain, from research institutes and breeders to processors, retailers and consumers.

## Urban and Peri-urban Agriculture

In the context of nutrition transition from fresh and nutritious foods to cheap energy-dense foods which are high in sugars, saturated fats and salts, including ultra-processed foods within and around cities, urban and peri-urban agriculture offers an opportunity to increase the availability of fresh and nutritious foods in proximity markets, and access to a diversified and nutritious diet for urban residents.

Urban and peri-urban agriculture can be integrated into the urban economy and embedded in the urban ecosystem, making use of available resources such as vacant and unused lands for urban farms, organic waste for compost and urban wastewater for irrigation, in accordance with Good Agricultural Practice and with a view to ensuring food safety. Commercial viability of urban and peri-urban agriculture can be increased by building capacities of producers and their organizations, and ensuring enterprise development through access to finance and markets.

## 2. Food Handling, Storage and Processing

### Nutrition-Sensitive Post-harvest Handling, Storage and Processing

Post-harvest handling, processing and storage contribute to: maintaining a secure supply of food (and thus of nutrients) throughout the year; preserving the quality of harvested raw material as it moves along the food supply chain from the producer to the market; reducing losses; and making fresh produce available in local markets as well as in distant locations. Food storage helps to maintain food quality over an extended period until its final use, permits its deferred use (on an annual or multiannual basis), guarantees the regular and continuous supply of raw materials for processing and helps to balance the supply and demand of agricultural products, thereby stabilizing market prices.

Strengthening the capacity of smallholders and small entrepreneurs, in particular women, to store, preserve, process and package foods can help secure a year-round food supply that can improve nutrition and income generation. This can be done through – for example –trainings on techniques to optimize the shelf-life and nutritional quality of foods. In addition to training, provision and maintenance of necessary equipment for storage (*e.g.* small silos), processing and packaging as well as sustainable supply of inputs, are key. Working with the food industry to improve or reformulate food composition of processed foods to reduce or eliminate the use of ingredients such as salt, trans fats, sugar and additives, is imperative.

### Food Fortification

While it is important to continue efforts to increase dietary diversity and quality as a long-term solution to all forms of malnutrition, food fortification policies can help to tackle micronutrient deficiencies through increasing the micronutrient content of staples and condiments consumed by large segments of the population and/or by vulnerable groups, such as children.

Fortification programmes can be mandatory and implemented on national scale via mass fortification (which is the preferred approach when the majority of the population is at risk of micronutrient deficiency) or voluntary, whereby the decision to fortify is taken by food manufacturers within the regulatory limits set by the government (*e.g.* fortification of porridge and other complementary foods for infant feeding). Fortification can also target specific vulnerable groups (*e.g.* provision of fortified school meals, emergency food distributions, or social protection programmes). Consumer demand for fortified products can be stimulated through social marketing and information campaigns, closely linked with nutrition education programmes.

## 3. Food Trade and Marketing

### Trade for Nutrition

Trade helps balance food deficits and surpluses across countries, facilitating the availability of food and contributing to price stability. By integrating national and international food markets, trade can help absorb domestic supply and demand shocks that could otherwise result in excess domestic food price volatility. In a context of globalization, urbanization and increased market reliance, looking at trade through a nutrition lens is increasingly important for maximizing benefits and reducing risks.

Lowering trade barriers for fruits and vegetables has great potential to improve nutrition through increasing their availability in importing countries, especially in counter-seasonal periods. Domestic trade, rural-urban linkages, short food supply chains where feasible, and city-region food systems should be promoted as a way to simultaneously increase access to fresh foods for consumers and to remunerative markets for producers.

## Food Marketing and Advertising Practices

Understanding the impact of marketing and advertising on consumer preferences, eating habits, diets and nutrition is crucial to design policies and strategies for shaping healthy food environments and leveraging both traditional and modern retail sectors to facilitate consumption of healthy diets.

Strategies to improve the retail environment need to acknowledge that supermarkets play a growing role and that they have a potential to affect nutrition both positively (via increased availability, accessibility and affordability of diverse and fresh foods) and negatively (by encouraging consumption of energy-dense, nutrient-poor and highly processed foods). A combination of incentives and regulations can be used to improve the nutrient value of products sold by retailers and caterers. For example: establishing nutrition standards for public procurement and catering services to ensure that food and menus supplied in schools, hospitals and public sector bodies are healthy and nutritionally balanced. This can also create incentives for farmers and the manufacturing sector to invest in nutritious foods.

## Food Price Policies for Promoting Healthy Diets

Food price policies have the potential to create incentives for increasing the supply and demand for nutritious food products, especially among vulnerable populations who cannot afford or adopt healthy food choices. Subsidies for selected nutritious foods can increase affordability of healthy diets and/or incentivize purchase. In particular, countries affected by the double burden of malnutrition should carefully select the right types of price interventions, considering that a given change in prices can positively impact some forms of malnutrition or population groups (*e.g.* reduction of undernutrition) and be harmful to others (*e.g.* increase in overweight and obesity).

## Food Labelling

Food labelling influences food choices by informing consumers on ingredients, health, safety, nutrition claims and nutrient content of a given item. Increasingly, labels inform consumers about foods with healthier nutrition profiles and can also motivate manufacturers to produce foods with healthier nutrition profiles.

Mandatory minimum requirements for food labelling are important to protect the consumer and contribute to safe, nutritionally balanced diets. Nutrition-related mandatory requirements typically include the ingredient list and the nutrition facts declaration (*e.g.* energy value, proteins, total fats, saturated fats, sugar) expressed as a percentage of reference values based on scientific data regarding nutrient requirements and dietary risks. Food advertising and marketing are closely related to labelling, as they can affect consumers' interpretations of the label, either reinforcing or detracting from the label message.

## 4. Consumer demand, Food Preparation and Preferences

### Nutrition Education and Behaviour Change Communication

Increasingly recognized as an essential catalyst for the success of food security and nutrition interventions, effective nutrition education ensures that increased food production/income translates into improved diets and improved nutrition status. Nutrition education is also becoming critically necessary in countries experiencing a dangerous dietary transition to cheap processed foods rich in sugar, fat and salt.

Nutrition education and behaviour change communication to consumers can be delivered through multiple venues and activities, and may include health and nutrition counselling during pregnancy, education on breastfeeding or improved complementary feeding of children under two years of age, nutrition education in schools and hands-on learning to enable families to practise good nutrition behaviours. Incorporating nutrition education into agriculture projects is likely to improve consumption and related nutrition impacts in the producing household. It can also incentivize households to diversify their production and retain more food for their own consumption. Agriculture and health extension workers can play an important role in facilitating community health and nutrition sessions.

### Income Generation for Nutrition

Ensuring regular and decent incomes for consumers is essential to achieve good nutrition, as it allows not only purchase of healthy foods, but also access to health care and education services. With increasingly commercialized agriculture and food systems, income becomes more important for nutrition than own food production. It thus becomes important to integrate nutrition objectives and components into interventions whose primary objective is to generate income (*e.g.* agriculture commercialization programmes).

Agricultural production and other food system activities offer opportunities for generating income through sale of agricultural products or wage labour, and can thus contribute to nutrition both directly through food production and indirectly through increasing economic resources, which can be used to buy nutritious foods. Targeting food crops that are both high in market value and nutritious, such as horticultural crops and milk, dairy and fish products, can have multiple benefits, as it increases availability of nutritious foods on-farm and in the market while increasing smallholder incomes. Value addition (through storage, processing, marketing, *etc.*) is key to generating income throughout the food system. Processing methods that preserve or add nutritional value should be encouraged to promote availability of nutritious foods on markets (*e.g.* dried mangos may be preferable to candy or biscuit making).

## Nutrition-Sensitive Social Protection

Safety nets and social protection schemes can play an important role in improving nutrition and addressing the social and economic determinants of malnutrition. Moreover, social protection helps families to increase their consumption and to access more and better food, while also helping to develop their productive asset base which is critical for sustaining good nutrition in the long term and facilitating access to health care and services.

Social assistance schemes designed to support the nutrition of vulnerable groups can thus take a variety of forms, including: In-kind transfers through general food distribution or targeted distribution of specialized foods for women and children (*e.g.* supplements). Nutrition impact of food transfers can be maximized by adding a nutrition education component, and by ensuring high nutritional quality of food baskets. This can include provision of nutrient-rich foods (*e.g.* animal-source foods, fruits and vegetables), fortified flours and biofortified staples. Social insurance schemes, including health insurance, targeted weather-based insurance for crops and livestock, maternity protection and employment insurance also contribute to protecting basic pro-nutrition assets against shocks and crises.

## School Good Nutrition

Good nutrition is key to children's physical and mental development. School nutrition is also important to reach adolescent girls – *i.e.* future mothers. By encouraging children, their families and the school community to develop life-long healthy eating habits, these programmes also contribute to generating new demand for, and supply of, healthy and nutritious foods.

School meals (*i.e.* cooked meals, snacks and take-home rations, as appropriate) must be based on children's dietary requirements and comprise a diversity of foods, including local foods for healthy, traditional diets. Beyond increasing children's access to basic and diverse foods, school meals can integrate fortified and biofortified food products. Schools can be used as platforms for delivering nutrition-specific interventions (*e.g.* deworming).

## Nutrition-sensitive Humanitarian Food Assistance

Humanitarian food assistance should ensure a balanced dietary intake for people affected by crises, and prevent acute and chronic malnutrition as well as micronutrient deficiencies. Moreover, it is important that interventions recognize and address the nutritional needs of specific groups and vulnerable members of the households (*e.g.* infants and young children, pregnant and lactating women, disabled and elderly people).

During emergencies, interventions should go beyond the delivery of food staples by providing access to a diversified diet containing the right foods and right nutrients, taking into account local preferences and eating habits. This could be done, for example, through provision of fresh food vouchers to complement household food baskets. During a crisis, the occurrence and spread of diseases increases, while access to health services, drinkable water, safe and hygienic living environments is undermined. Nutrition-sensitive humanitarian assistance should seek to integrate with health and water, sanitation and hygiene-based interventions, so as to tackle the different determinants of malnutrition and to maximize impact.

## Cross-cutting Issues

### Nutrition-Sensitive Value Chains

Market linkages and value chains play a critical role in determining availability, affordability and quality of food. In this context, nutrition-sensitive value chain approaches offer a useful framework to navigate the complexity of the food system and to unleash its potential to deliver healthy foods, by maximizing nutrition opportunities at any step of the value chain. Value chain analysis with a nutrition lens can help to re-design existing value chain interventions to achieve greater nutritional impacts (*e.g.* through assessing the nutrition implications of an existing value chain and introducing appropriate changes in the way the chain is organized). Promoting investments in value chain logistics is an integral part of trade strategies and policies. In particular, value chains can be leveraged to improve rural-urban trade linkages and to achieve a win-win outcome in which rural producers enjoy greater economic returns and urban consumers enjoy a diversity of nutritious foods at affordable prices.

### Women's Empowerment and Gender Equality

Gender-based inequalities between men and women have a strong impact on the population's nutritional status when women do not have access to family income or other resources (land, credit, information, *etc.*) or are not empowered to make decisions on their use and distribution. Furthermore, women's workload (in fields, fetching water and fuelwood, domestic chores, *etc.*) can result in reduced time for childcare, breastfeeding and food preparation. Designing and implementing gender-sensitive interventions in agricultural and rural development and the food system, which address unequal gender relations and empower women, are major factors contributing to the success of programmes to improve nutrition.

Women play a key role in all phases of the value chain including reducing food loss and waste at the production, post-harvest and processing stages. Interventions that focus on enhancing women's knowledge and capacities are therefore important to ensure that an adequate quantity of food is available and that the nutritional value of food is preserved. Involving all members of the household as well as community leaders, who often play an important role in household dynamics and decision-making, is an important factor for successfully ensuring sustainable behavioural change and achieving better nutrition and recognition of women's key roles. In particular, efforts should be made to involve fathers and grandparents and to build their knowledge on nutrition and care. Fathers should be encouraged to participate actively and share responsibilities with mothers in caring for their infants and young children.

## Food Loss and Waste: Prevention, Reduction and Management

Food loss is the decrease in quantity or quality of safe and nutritious food available and accessible for direct human consumption. Food waste is an element of food loss and refers to discarding or making an alternative (non-food) use of safe and nutritious food at any point along food supply chains. Causes of food loss and waste are context-specific and may relate to gaps in capacity of the food supply chain actors, inadequate storage facilities and food packaging, lack of access to markets and consumer behaviour. While most waste occurs in industrialized countries, it is also becoming a growing issue for developing countries, due to such factors as urbanization, gaps in the urban-rural continuum, inefficient distribution chains and changes in diets and lifestyles.

In developing countries food losses occur mainly at early stages of the food value chain. Strengthening the supply chain through the direct support of farmers and investments in cold chain infrastructure, transportation, and safe packaging could help to reduce the amount of food loss. Food waste prevention at its source can be achieved through recovery and redistribution of safe and nutritious food for direct human consumption. Recovery involves receiving, with or without payment, food that would otherwise be discarded or wasted. Redistribution means to store or process and then distribute the received food directly or through intermediaries, with or without payment, to those in need.

## Food Quality, Safety and Hygiene

Food contaminated with biological, chemical or physical hazards, including harmful pathogens, natural toxins and chemicals can contribute to undernutrition and cause adverse health effects. Food quality, hygiene and safety standards are therefore systematic preventive approaches to food safety that aim to protect public health and improve accessibility of nutritious and safe foods in ways that address modern food environments.

Risks can be controlled at various points of the supply chain, including through: reduction of pesticide use in cultivation and antibiotics in animal production; prevention of harvest contamination by animals; implementation of basic sanitation; air circulation and humidity controls in storage and processing facilities; aflatoxin control; improved hygiene and safety practices of street food vendors; and delivery of messages to households on hand-washing and safe food handling and preparation. Where the informal sector predominates, it is advisable to "professionalize" rather than "penalize". Combining capacity development of the informal sector with incentives to further motivate behaviour change has proven an effective approach to advancing food safety in many developing countries.

# Recommendation and Enabling Environment for NSA and Food Systems

Agriculture programmes and investments need to be supported by an enabling policy environment if they are to contribute to improving nutrition. Governments can encourage improvements in nutrition through agriculture by taking into consideration the following policy actions below:

✩ Agricultural policies need to support the production of a diversity of nutrient-rich foods and to be congruent with national nutrition priorities and goals, including national food-based dietary guidelines.

✩ Adoption of sectoral and cross-sectoral frameworks and approaches – including for crops, livestock, forestry, fisheries and aquaculture – will facilitate transition to more sustainable and diverse production systems.

✩ The potential of livestock and fishery sector to contribute to address malnutrition problems is high, but still undervalued. Building a body of evidence in nutrition-sensitive livestock and fishery, and developing capacity in integrated nutrition and livestock/fishery programming is necessary to foster a 'nutrition-sensitive' culture among experts of animal-based interventions and to ensure that these interventions are included in the portfolio of dietary diversification strategies supported by nutrition experts.

✩ In a context where most subsidies, investments and research programmes are concentrated on major staple grains and selected animal species, policy support for other foods (including fruits, vegetables, pulses and underutilized species) is key to realize the full potential of nutrition-sensitive agriculture and biodiversity to improve nutrition and health. Regulatory mechanisms for protection of biodiversity in highly competitive markets should be developed, as elements of broader policies for tackling ecosystem degradation.

✩ Public sector investment to strengthen national agriculture research and extension systems and seed producers is crucial to ensure continuous production of high quality, nutritious seeds. Regulatory and legal frameworks which provide harmonized standards for claims regarding quality, nutrient levels, health benefits and biosafety of biofortified crops need to be developed at international and national levels.

✩ Policies in support of urban and peri-urban agriculture recognize urban food production as a legitimate economic activity, enhance access to vacant lands (with clean soil and water) and integrate urban agriculture in land-use planning.

✩ A good infrastructural support base, including efficient cold chain infrastructure, is required to facilitate and support post-harvest handling operations. Different kinds of instruments can be used to promote nutrition-sensitive food reformulation at manufacturer levels, including:

❑ Incentives (*e.g.* funding allocated to schools that choose processed products free of added sugar, fats and salt, as "healthy snacks");

❑ Voluntary and co-regulatory schemes (*e.g.* salt reduction initiatives) entailing agreements between governmental bodies and the private sector, including manufacturers and retailers, with the oversight of public health experts from research institutes and hospitals;

☐ Mandatory approaches (*e.g.* mandatory labels indicating high salt content, mandatory limits on levels of salt, bans on trans fats in food products).

☆ Legal frameworks are needed to set mandatory fortification and appropriate technical standards (on the basis of WHO guidelines), as well as mechanisms for monitoring and controlling quality of fortified products and truthfulness of nutrition claims. National legal frameworks, based on relevant international (*e.g.* WHO) standards, should prevent inappropriate promotion of fortified food for infants and children and ensure that it does not undermine support to optimal breastfeeding practices and use of locally available and affordable nutritious foods for complementary feeding.

☆ Adoption of harmonized standards (*e.g.* sanitary and phytosanitary, as well as food labelling standards based on the Codex Alimentarius) is an option for facilitating trade while ensuring food safety and protecting consumer health. However, consideration should be given to small farmers and small-scale food processors (especially in the informal sector), who might not be able to comply with such standards, in order to avoid their exclusion from the market.

☆ Enabling farmers' markets, small shops and traditional retailers and caterers to survive and compete on a level playing field with large players helps to create a more balanced retail environment where the healthy choice becomes the easy choice (*e.g.* purchasing the right quantity of foods, choosing fresh over ultra-processed items).

☆ Adoption and enforcement of competition policies in the food sector, which is often prone to collusion and other forms of anti-competitive practices, are fundamental to ensure fair prices for both producers and consumers.

☆ Policies regulating nutrition labelling and claims, if carefully designed, can have positive impacts on nutrition by promoting foods with healthy profiles attractive for consumers. Nutrition labelling policies should take into account consumer use, interpretation and understanding of different nutrition labelling schemes.

☆ National food-based dietary guidelines should form the basis for public food and nutrition, health and agricultural policies and nutrition education programmes to foster healthy eating habits and lifestyle. Education curricula and linkages with universities and learning institutions should be set up to ensure proper professional training for nutrition educators.

☆ Legal frameworks to support decent employment in the formal sector, as well as innovative approaches to facilitate employment in the informal sector, are key elements of the enabling environment for income generation for nutrition.

☆ Social protection is more likely to be nutrition-sensitive if it is part of an integrated package of sectoral policies including food assistance policies, health policies (*e.g.* universal health coverage with special focus on ante-

and post-natal care, pediatric services and immunization), labor regulation (*e.g.* maternity protection), gender equality and women's empowerment policies, among others.

☆ Policies and regulatory frameworks should include measures for healthy school environments – *e.g.* to ensure that food available and sold in school environments is of high nutritional quality, to protect children from marketing of unhealthy foods, to ensure sustainable and equitable access to and use of safe water and basic sanitation services. Effective integration of school food and nutrition into national programmes and policy frameworks is key to ensure effectiveness, sustainability and stable funding.

☆ In contexts of emergencies and protracted crisis, national capacity building on nutrition should also enhance skills of those involved in the design and delivery of humanitarian food assistance programmes. Addressing the disconnect between the food security and the nutrition sectors during emergencies – and encouraging a coordinated approach – is crucial to ensure cohesive programming and a community of good practices.

☆ Value chain for nutrition approaches are often implemented in the context of public-private collaboration. However, transparent and inclusive policy frameworks are needed to manage potential risks and trade-offs between (private) economic objectives and (public) health goals.

☆ Health and nutrition policies and guidelines should pay specific attention to the physiological needs of pregnant and lactating women, addressing their greater vulnerability to malnutrition and micronutrient deficiencies. Promoting and facilitating women's access to education, including high school and higher education is a fundamental foundation for women's empowerment.

☆ Policies to address food loss and waste need to be established from production to consumption, supported with adequate budget allocation, monitoring and evaluation frameworks, and coordinated both vertically (from national to local levels) and horizontally (between sectors). Enhanced availability of data on food loss and waste is an important element of an enabling environment for informed decisions and multistakeholder collaborations, including for achieving SDG 12.3.

☆ National policies on food quality, safety and hygiene should be carefully designed so as to avoid unintended discrimination against small and medium-sized enterprises, which constitute the backbone of economies in most developing countries, and informal markets, from which poor consumers often source much of their food. It can be necessary to adapt regulations to specific needs of these enterprises in order to reduce the regulatory burden without compromising consumer health and safety.

# REFERENCES

FAO. 2014. *Nutrition-Sensitive Agriculture.* Retrieved August 12, 2020, from Food and Agriculture Organization of the United Nations (FAO): http: //www.fao.org/3/a-as601e.pdf

FAO. 2017. *Nutrition-sensitive agriculture and food systems in practice.* Retrieved August 12, 2020, from Food and Agriculture Organization of the United Nations (FAO): http: //www.fao.org/3/a-i7848e.pdf.

Chapter 3

# Creating Sustainable Value Chains for Food System Transformation in South Asia



**Nafees Meah**

*IRRI Regional Representative for South Asia*



## Introduction

The South Asia Region, comprising Afghanistan, Bangladesh, Bhutan, India, Nepal, Pakistan, and Sri Lanka, is home to ca. 1.7 billion people with a combined GDP in 2018 of 3.45 trillion USD (World Bank Development Indicators 2019). More than two-thirds of the population in the region still live in rural areas and depend on agriculture, at least in part, for their livelihoods. The Green Revolution helped transform the South Asia region from one of food deficits to surpluses and moved millions of people out of poverty (Pingali 2012). However, despite rapid economic growth experienced across the region since the 1990s, a very high incidence of poverty, hunger, and malnutrition persists - particularly in rural areas.

Smallholder farmers provide ca. 80 per cent of the food supply in Africa and Asia (IFAD 2013). In addition, in many countries in South Asia, smallholder farmers are transforming from subsistence to commercial agriculture (Reardon *et al.*, 2013). If over the next decades smallholder farmers can improve on-farm productivity, increase resource-use efficiency, diversify their crops, gain better market access and find non-farm employment for their households, then the livelihoods, nutrition, and incomes of millions of smallholder farmers and their families could be improved substantially.

All South Asian countries are also signatories to the UN's Sustainable Development Goals (United Nations 2015) and have adopted a set of goals to end poverty, protect the planet, and ensure prosperity for all by 2030. However, rising populations, increasing economic inequality, high rates of poverty, large malnourished population, a big gender gap, environmental challenges from climate change, land degradation, water stress *etc.* mean that achieving these for South Asia will need concerted and sustained effort.

## A Renewed Focus on Food Systems

Countries in South Asia are dealing with the three issues of undernutrition, micronutrient deficiencies and overweight/obesity at the same time ("triple burden') (Meenakshi, 2016). Public Health Foundation of India reported that stunting in children aged under 5 years of age in states ranged from over 50 per cent (Uttar Pradesh) to 19 per cent (Kerala) (Raykar *et al.*, 2015). In Sri Lanka, 15 per cent of under 5-year olds are stunted (UNICEF 2018). In Bangladesh, among the children under 5 years, 45 per cent are zinc deficient, 11 per cent are iron deficient, and 21 per cent are vitamin A deficient (Ahmed *et al.*, 2016). Headey and Hoddinott (2016) in a study on agriculture, nutrition and the green revolution in Bangladesh found that although the country achieved rapid growth in rice productivity and achieved food security in terms of meeting calorie needs, there was relatively sluggish diversification in food production and consumption. There was no observed improvement in the rate of stunting in children (Headey and Hoddinott, 2016).

Globally, the food system has also been adapting to rapid population growth over the last few decades. However, more than 800 million people still have inadequate access to food (FAO, IFAD, UNICEF, WFP and WHO 2019). In addition, a growing share of the world population suffers from micronutrient deficiencies or is overweight or obese, leading to an increasing prevalence of non-communicable diseases. It is now widely accepted that the existing food system has failed and the problem needs to be addressed through evolving and improved food systems (UN Environment 2018).

## By a Food System we mean

A consensus has yet to emerge on the definition of a food system. In this paper, we use the definition by the Oxford Martin Programme on the Future of Food found at *https://www.futureoffood.ox.ac.uk/what-food-system ...a complex web of activities involving the production, processing, transport, and consumption. Issues concerning the food system include the governance and economics of food production, its sustainability, the degree to which we waste food, how food production affects the natural environment and the impact of food on individual and population health.* Food systems are at the nexus of food security, nutritional health, viable ecosystems, climate change, and prosperity. Until now national agricultural policies in developing countries have tended to focus on increasing food production. In most cases, these policies have neglected the other roles and responsibilities of agriculture within food systems. Conventional agricultural policies do not address negative externalities on nutritional health, natural capital, and protection of biodiversity (Benton and Bailey 2019). FAO have estimated that the natural capital costs associated with crop production was 1.15 trillion USD globally *i.e.* over 170 per cent of its production value (FAO 2015). It is recognized that agriculture and food policies should align with the 2030 Agenda for

Sustainable Development by addressing several human- and planet-centred goals simultaneously (United Nations 2015). The EAT Lancet Commission, in developing the concept of *planetary boundaries*, proposed a "safe operating space" for food systems based on existing scientific evidence for healthy diets and sustainable food production (Raworth 2012; EAT Lancet Commission 2019). Operating outside this space would, they argue, increase the risk of harm to the stability of the earth system processes (such as high rates of biodiversity loss) and human health. They argue that the "scientific targets that define the safe operating space for food systems allow the evaluation of which diets and food production practices will enable achievement of the SDGs and the Paris Agreement".

Sustainable development, more generally, is often presented with three parts: social, economic and environmental. These are asserted to be of equal importance to the overall sustainability of the system. This is encapsulated in the classic definition presented by the Brundtland Commission that "Sustainable development meets the needs of the present without compromising the ability of future generations to meet their needs". However, one should not assume that there will always be *win-win* situations between the different elements of food system sustainability. In reality, there will be *trade-offs* to be made between the different elements of sustainability in a given concrete situation. That is to say one may need to choose between how much social or economic or environmental elements of sustainability to have and the timing and sequencing of these. Indeed, *win-win* situations may be rare or difficult to implement and, therefore, more often than not, some difficult societal choices will be need to be made (Béné *et al.*, 2019).

## Sustainable Food Value Chains

An effective food system transformation will necessitate firm commitments at the international and national levels to foster the required policies and investments at national and local levels. In practical terms, this would need to include four main dimensions: (i) underpinning healthier populations by enabling access to nutritious and healthy food for all; (ii) guaranteeing sustainable food production, processing, trade, and retailing; (iii) mitigating and adapting to climate change; (iv) improving smallholder farmer livelihoods and resilience by enhancing prosperity in farming and rural communities.

One contribution to operationalising the food system transformation is to develop sustainable food value chains for appropriate cropping and mixed livestock/fish systems *tailored* for specific regions and agro-ecologies (Figure 3.1). The idea is to strengthen the value chain so that the objectives of higher farmgate prices and better quality of food products for consumers is achieved. As part of a burgeoning literature on food value chains, the FAO has developed guiding principles on sustainable food value chains which can be used as a framework for upgrading value chains (FAO 2014b).

## By a Food Value Chain, we mean

*The full range of farms and firms and their successive coordinated value-adding activities that produce particular raw agricultural materials and transform them into*

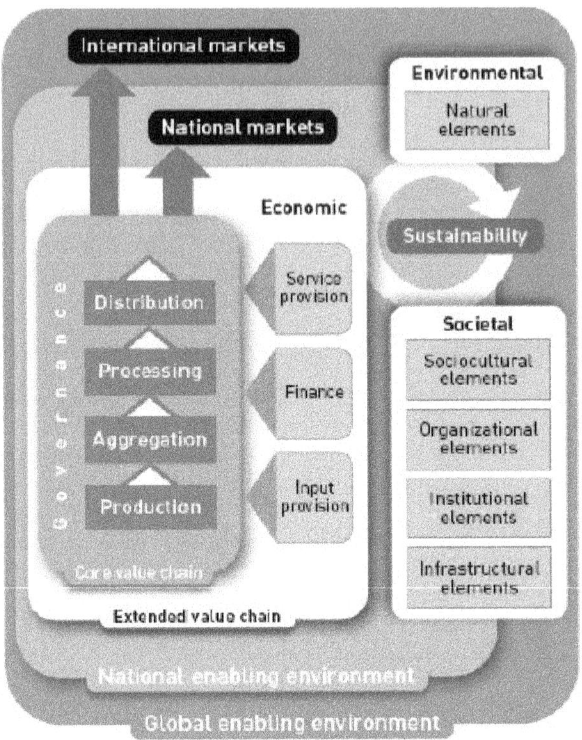


**Figure 3.1**

*particular food products that are sold to final consumers and disposed off after use, in a manner that is profitable throughout, has broad-based benefits for society and does not permanently deplete natural resources. (FAO 2014b).*

A value chain approach would engage the private sector actors and ensure that the focus is on increasing the overall productivity, aggregate profitability and efficiency of food production. Smallholder farmers in South Asia would be the overall beneficiaries provided that there is the necessary *trust* between the value chain actors (based on the principle that everyone receives their *fair share*) and there is a clear market demand-led process. Given that the vast majority of farmers in South Asia are smallholders, this entails some horizontal coordination between producers *i.e.* some kind of Farmer Producer Organizations (FPOs) or looser associations as well as vertical coordination (FAO 2014b). Failing that, there is the risk that other actors in the value chain will assume a dominant position in a particular value chain. Work has shown that rice value chains in India and Bangladesh are changing rapidly. Farmers are selling directly to millers, by-passing village traders – *disintermediation* – and thereby retaining more of the value (Reardon *et al.*, 2013). However, in some food value chains modern food industry firms (*e.g.* supermarkets) are making major inroads in South Asia (and especially in South East Asia) and, in order to ensure consistent and higher quality of agricultural products, are entering into contract farming arrangements (Ba *et al.*, 2019). The evidence shows that contracts as part of this vertical integration vary from those where farmers are provided with most inputs (*e.g.* seeds, machinery) and training to those where they are contracted simply to provide raw agricultural products (FAO 2014b). As regards the improving nutrition, this could be addressed through meeting the market demand for fruits and vegetables. Whether primary producers diversify their own diets and, thereby, improve their nutritional status is a more complex issue. Gillespie *et al.*, reporting on the results of the LANSA (Leveraging agriculture for nutrition in South Asia) initiative have shown that the various pathways from agriculture, as a source of food production and livelihood, to nutrition do not operate in isolation but they interact. Thus, for example, if women's role in agriculture as labour increases, but this is not accompanied by an increase in decision making power within the family, then the nutritional outcomes for them and their children may become worse (Gillespie *et al.*, 2019). This calls for, not only more research, but also for more nuanced interventions by public authorities aiming to improve nutritional and health outcomes.

In respect of environmental sustainability, the value chain should seek to create additional economic value without adversely impacting natural capital. There may be a need to recognise and negotiate trade-offs between the actors in the value chain as a more environmentally-friendly value chain products may not generate the additional income for farmers. However, there is increasing evidence that in the developing economies, consumers are willing to pay a premium for environmentally-friendly food products (Ba *et al.*, 2019). For the rice sector, the Sustainable Rice Platform, a global alliance of agricultural research institutions, agri-food businesses, and public sector and civil society organisations has been established by United Nations Environment and IRRI (SRP 2017). This has developed standards and performance indicators for sustainable rice cultivation covering social, economic and environmental dimensions of rice production that contribute to an increased global supply of affordable rice, improved livelihoods for producers and reduced environmental impact of production.

## Conclusions

The Green Revolution helped transform the South Asia region from one of food deficits to surpluses and moved millions of people out of poverty. Despite rapid economic growth experienced across the South Asia region since the 1990s, a very high incidence of poverty, hunger, and malnutrition persists - particularly in rural areas. It is now widely accepted that the existing food system has failed and the problem needs to be addressed through evolving and improved food systems. In South Asia, one contribution to operationalising the food system transformation is to develop sustainable and inclusive food value chains for appropriate cropping and mixed livestock/fish systems *tailored* for specific regions and agroecologies. For this to happen, there will need to be policy coherence and convergence at local, national and global levels and articulation of the inevitable trade-offs between nutritional health, environmental sustainability, economic prosperity and food security.

# REFERENCES


Ahmed F, Prendiville N and Narayan A. 2016. Micronutrient deficiencies among children and women in Bangladesh: progress and challenges. *Journal of Nutritional Science*, vol. 5, e46, 1-12

Ba H, de Meyb Y, Thorond S and Matty Demont. 2019. Inclusiveness of contract farming along the vertical coordination continuum: Evidence from the Vietnamese rice sector. *Land Use Policy*, 87, 104050

Béné C, Oosterveer P, Lamotte L, Brouwer I, de Haan S, Prager S, Talsma E, Khoury C. 2019. When food systems meet sustainability – Current narratives and implications for actions. *World Development*, 113, 116–130.

Benton T and Bailey R. 2019. The paradox of productivity: agricultural productivity promotes food system inefficiency. *Global Sustainability.* 2, e6, 1–8

Brundtland Commission *Our Common Future 1987*. EAT-Lancet Commission (2019) The EAT-Lancet Commission on Food, Planet, Health. Retrieved 16th September 2019.https://eatforum.org/eat-lancet-commission/

FAO. 2014a. *Regional Rice Strategy for Sustainable Food Security in Asia and the Pacific.* FAO. Bangkok

FAO. 2014b. *Developing sustainable food value chains – Guiding principles.* FAO.Rome

FAO (2015) *Natural Capital Impacts in Agriculture.* FAO. Rome

FAO, IFAD, UNICEF, WFP and WHO. 2019. The State of Food Security and Nutrition in the World 2019. Safeguarding against economic slowdowns and downturns. Rome, FAO

Gillespie S, Poole N, van den Bold M, Bhavani R, Dangourd A and Shetty P. 2019. Leveraging agriculture for nutrition in South Asia: What do we know, and what have we learned? *Food Policy, 82, 3–12*

Headey D and Hoddinott J. 2016. Agriculture, nutrition and the green revolution in Bangladesh. *Agricultural Systems*, 149, 122 - 131

IFAD.2013. Smallholders, food security, and the environment. IFAD Rome

Meenakshi JV. 2016. Trends and patterns in the triple burden of malnutrition in India. Centre for Development Economics. Delhi School of Economics. *Working Paper No.* 256

Mellor, J. W. 2017. The Economic Transformation. In *Agricultural Development and Economic Transformation*. Springer International Publishing AG. Cham. Switzerland.

Pingali P. 2012. *Green Revolution: Impacts, limits, and the path ahead.* PNAS July 31, 109 (31) 12302-12308

Raykar N., Majumdar M, Laxminarayan R., Menon P. 2015. *India Health Report: Nutrition 2015.* New Delhi, India: Public Health Foundation of India.

Raworth K. 2012. *A safe and just operating space for humanity. Can we live within the doughnut?* Oxford: Oxfam Discussion Papers

Reardon T, Minten B, Chen K and Lourdes A. 2013. The Transformation of Rice Value Chains in Bangladesh and India: Implications for Food Security, *ADB Economics Working Paper Series, No. 375, Asian Development Bank.* Manila

The Economics of Ecosystems and Biodiversity (TEEB). 2018. Measuring what matters in agriculture and food systems: a synthesis of the results and recommendations of TEEB for Agriculture and Food's Scientific and Economic Foundations report. Geneva: UN Environment.

The Sustainable Rice Platform (SRP).2017.Retrieved 16th September 2019. http: // www.sustainablerice.org/

UNICEF. 2018. Child Stunting, Hidden Hunger and Human Capital in South Asia: Implications for Sustainable Development Post 2015. UNICEF: Kathmandu, Nepal

United Nations (2015) "Transforming our world: the 2030 Agenda for Sustainable Development". *United Nations – Sustainable Development knowledge platform.* Retrieved 16th September 2019.https: //sustainabledevelopment.un.org/ post2015/transformingourworld.

World Bank. 2019. *World Bank Development Indicators, 2019.* Retrieved 16th September 2019. *http: //datatopics. worldbank.org/world-development-indicators/*

# Chapter 4

# Sustainable Agriculture Practices for Food Security and Zero Hunger in India

**Ch. Srinivasa Rao**

*Director,*
*ICAR-NAARM, Hyderabad*

United Nations Member States in 2015 adopted a universal call to end poverty, protect the planet and ensure that all people enjoy peace and prosperity by 2030 by observing Sustainable Development Goals (SDGs), also known as the Global Goals. Year 2017 marks the first year of implementation of the United Nations SDGs. The SDGs are a set of 17 global goals measured by progress against 169 targets covering social issues like poverty, hunger, health, education, climate change, gender equality, and social justice. Before SDGs, United Nations in year 2000 adopted eight international development goals referred as Millennium Development Goals. The goal 2nd of SDGs refers to "Zero Hunger" and member nations are committed to have hunger free countries by 2030. Agriculture is critical in reaching SDGs and some 4-5 SDGs are directly linked to agriculture sector. Poverty and education are two principal factors that are linked to hunger.

Literacy is fundamental in observing and implementing SDGs as is also evidenced by the data available. The status of literacy in India is only 74.04 per cent which is far less than the global average of 86.2 per cent. It is little less in rural areas (68.91 per cent) and only 55 per cent in people with disabilities. The literacy rate in the country is 74.04 per cent, 82.14 for males and 65.46 for females. Kerala is on top with a 93.91 per cent literacy rate, closely followed by Lakshadweep (92.28 per cent) and Mizoram (91.58 per cent) where as Bihar with a literacy rate of 63.82 per cent ranks last in the country preceded by Arunachal Pradesh (66.95 per cent)

and Rajasthan (67.06 per cent). Kerala is comfortably good in implementing and observing SDGs that can be linked to literacy and similarly Bihar is at the bottom in observing SDGs.

## Indian Agriculture and SDGs Challenges

1. *Farm Holding:* Farm holding is a major constraint in observing mechanisation and adoption of technologies in agriculture production system. Over 80 per cent of farmers in India are marginal and small farmers with landholding of less than 2 hectares. Can India have aggregation of

farmers to have cooperative farming or collective farming remains an important question.

2. ***Degradation of natural resources and forest ecology***: 1/3ʳᵈ as forest cover is important to keep balance in ecosystem. Over use or miss use of natural resources has depleted natural resources and shall lead to reduced and nutrient deficient food production.

3. **Climate Change:** The change is visible. Weather pattern has changed, despite being no change in average rainfall, the rainy days have reduced. Some parts of the country are getting floods and some remain dry and draughted.

4. **Weak markets:** Marketing in the country is distrusted. Farmers or producers receive roughly 1/3ʳᵈ of the actual price and middle man gets the more. India is looking for one market and one nation. MSP need to be accurately worked out and effectively implemented. Insurance cover need to be offered in all the sectors of agriculture.

5. **Primary agriculture:** Indian agriculture is primary agriculture. What we produce, we mostly consume it and not spare it for processing. Developed countries process the foods and value adds it and earns high prices. Indian farmers are having low income as a result low educated and thus low quality life. India has witnessed many revolutions like green revolution, golden revolution, blue revolution, silver revolution, white revolution and these revolutions have increased food production to the extent of self sufficiency and India produces 290 million tons of grains, 320 million tons fruits and vegetables that are more than the sufficient. However, India need to produce more to effect export of food and earn foreign exchange and strengthen its economy.

6. **Biodiversity:** India is rich in resource biodiversity but little has been harvested. India has great varieties in plant, animal and marine species that can be exploited and production increased.

7. **Hunger Index:** Over 14 per cent of Indian population of which women constitute 60 per cent are going to bed hungry. India ranks at 102 position in Global Hunger Index among 117 countries close to many African countries despite having sufficient food production. The accessibility to food is the main challenge to end hunger in India. In Covid 19 pandemic, all the sectors in the country got down except the agriculture sector. People use to get groceries, milk, milk products, eggs even in strict lockdown situations.

8. **Natural Resources:** Land, water, soil, forest, climate, biodiversity are the natural resources that provide food security, nutritional security and of course health. Agriculture food production has witnessed continuous increase, however, observed many dips due to emerging pest diseases, water depletion, nutrient loss, energy scarcity, climate change, low utilization efficiency. You need to adopt new technologies, new varieties/breeds, stress tolerant cultivars/breeds to push the production curve up.

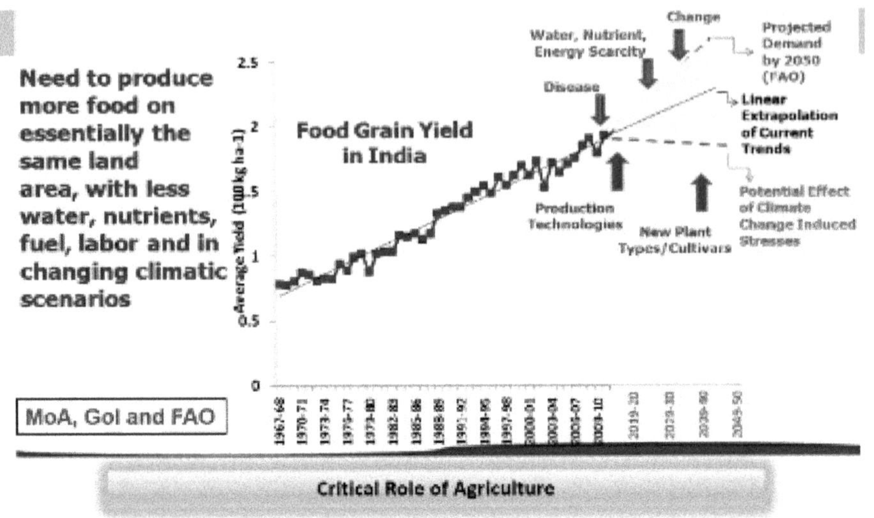


Sustainability does not mean to reduce production but increase the production with judicial use of resources and their conservation. Among the natural resources water, soil and climate are most important in agriculture production system.

a. **Water management:** It is the most important natural resource to produce food. Water crisis has remained the main reason for human migration. The water table has decreased significantly and the water withdrawal for food production has increased. It has resulted into increased energy expenditure on food production. More efficient technologies are needed to be employed in food production like drip irrigation and micro-irrigation system. India has adopted Pradhan Mantri Krishi Sinchai Yojanaa with aim to have more crop/food per drop of water. Farm pond technology is again an effective technology adopted in Hyderabad that has changed the agriculture in the State. Sea water farming is also a great opportunity with India to increase protein food. India has roughly 7500 Km costal line that can be effectively used for cage culture, Brackish water aquaculture to boost the protein availability.

b. **Soil conservation:** Soil is the gold and we need to give back to the soil what we have taken from it. Our soil is deficient in most of the nutrients be it carbon, nitrogen, potassium, phosphorus. Earlier we use to get 70 Kg grain per Kg of fertilizer and now only 3 Kg per Kg fertilizer. Crop residues are being burned depriving the soil of carbon and polluting the atmosphere. Soil cover is very important for soil enrichment and bare soil is very dangerous situation. Biochar use is an effective technology to enrich soil with carbon and water holding capacity. Nano-technologies like Nano fertilizers and micronutrients can reduce the dose of fertilizers but increase the utilization efficiency. Integrated farming can be a useful technology particularly in hill and Himalayan ecosystem. Diversified crop cultivation like vegetables, legumes, duck and fish culture along with paddy cultivation is integrated farming. It is somewhat labour intensive that can be solved with mechanisation.

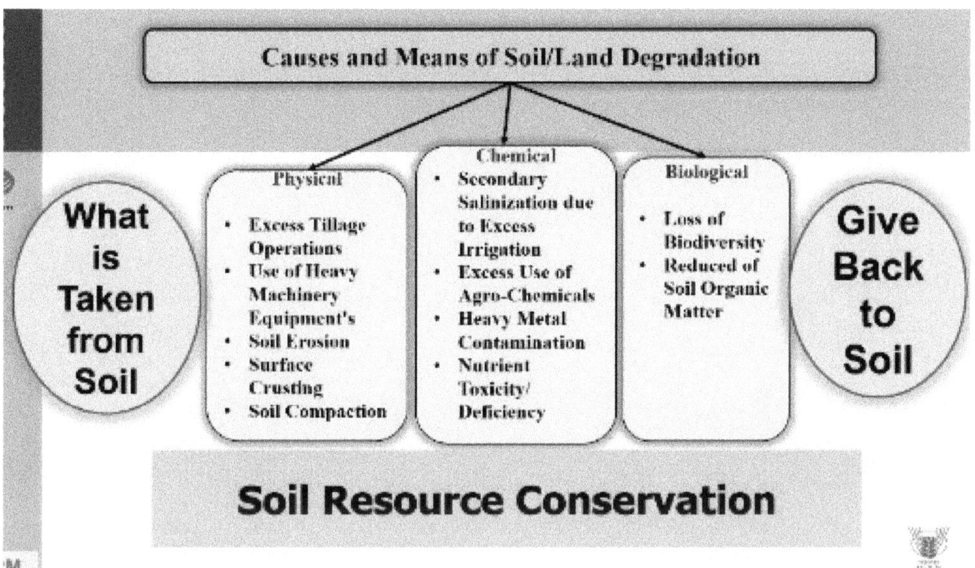


c. **Climate change:** India is more vulnerable to climate change. Most of the States are 50 per cent to 60 per cent where as Gujarat and Rajasthan are 100 per cent vulnerable to climate. India has released the policy paper "Climate Change and Indian Agriculture: Impacts, Coping Strategies, Programmes and Policies" that is available in public domain.

# Adaptation Strategies

- ☆ Draught/heat resistant varieties/breeds
- ☆ New farm management practices
- ☆ Change in land use
- ☆ Water management
- ☆ Agriculture insurance
- ☆ Agro-forestry system – location specific
- ☆ Resource conservation technologies
- ☆ Enriching soil organic matter
- ☆ Bio-fuels

( Chapter 5 )

# The Achievement of Food and Nutrition Security in South Asia is Deeply Gendered

**Nitya Roa**

*School of International Development at University of East Anglia (UEA)*



Women form an integral part of the agricultural sector, and in much of South Asia women make up a majority of the agricultural workforce and are often compelled to work to meet their families' basic needs. While their contributions are recognized as central to the food and nutrition security of households and communities, their work is not recognized or supported adequately by public policy and social institutions. Women continue to face inequality across key development indicators including health, education and nutrition; discriminatory laws; and high levels of precarity in terms of income, employment conditions, safety and well-being. Social structures that promote gender inequality and inhibit the agency of women contribute to the South Asian enigma — the persistence of under-nutrition despite economic growth — and must be addressed to achieve food and nutrition security.

Twenty-five years have passed since the Fourth World Conference on Women in Beijing. This was a landmark event for setting in place a comprehensive global policy framework on women's rights and gender equality and promoting feminist mobilization and alliance-building within countries and transnationally. The record of success over the intervening years has been equivocal. Progress has been made in education, life expectancy and political participation, including quotas for women in local councils, legislation on domestic violence and inheritance rights. Yet laws are rarely enforced, and gender inequalities have deepened across sectors in many parts of the world, including South Asia[1]. This Perspective posits that gender inequalities and the social structures that support them, specifically in the agri-food sector, underpin persistent food and nutrition insecurity in South Asia and deserve greater attention in terms of policy and law enforcement. The South Asian enigma despite economic growth, food and nutrition security has not improved in South Asia, a trend commonly referred to as the South Asian enigma. South Asia continues to be placed in the 'serious' category by the Global Hunger Index, with India performing worse than all of its neighbours, barring Afghanistan[2]. Poor child nutrition outcomes in South Asia are generally observed to result from proximal causes such as poor infant and young child feeding practices, poor nutrition among women before and during pregnancy, and poor sanitation practices[3,4]. However, there are multiple contextual drivers relating to gender that contribute to the South Asian enigma — the underlying social structures that support inadequate nutrition through the suppression of women's agency. In addition to the question of female survival itself, resulting from sex selection before birth, labour market arrangements that devalue women's work and perpetrate gender wage gaps[5], unequal access to productive assets[6] and growing incidents of violence against women[7] all contribute to this enigma. These factors vary across agro-ecologies and farming systems, and according to the social and economic status of households[8].

Violence and risk of food and nutrition insecurity: Violence against women is a structural barrier to the attainment of food and nutrition security via a range of pathways: withholding food or restricting funds to purchase food[9]; controlling when and how they eat[10], pushing women into high-risk behaviour to secure money for food[11] or the normalization of physical violence related to the non-performance of food-related work (food production, shopping, cooking and so on)[12,13]. Whereas food deprivation itself is a form of violence, it can at the same time aggravate other forms of violence, including early marriage and trafficking of young girls or intimate partner violence. Violence occurs throughout the life-course[14].

In India, despite legislation such as the Preconception and Prenatal Diagnostic Techniques Act passed in 1994 to prevent sex-selective abortions, child sex ratios continue to decline, from 927 girls to 1,000 boys in 2001 to 919 in 2011[15]. The growth in the Indian economy over this period has not helped reverse this trend, although if the girl child survives, she is likely to receive equal opportunities in terms of schooling, at least. This is partly a result of affirmative action, including the recent policy campaign to 'educate and save the girl child' (beti padhao, beti bachao). Once she reaches adulthood, different forms of domestic violence, including dowry-related harassment, again take centre stage, despite the existence of laws to protect women16. In fact, specific campaigns to empower women, including financial inclusion through self-help groups or support for secondary education of girls, seem to be evoking a violent backlash, intensifying varied forms of 'food violences' in the process[10,17]. With an incidence rate of roughly 40 per cent, the experience of physical violence is particularly severe for women from rural areas who belong to the Scheduled Castes and Tribes, who are also amongst the lowest wealth quintiles and least educated. In the absence of law enforcement, such forms of violence have consequences for well-being in general, and the nutritional status of women and their children in particular[18,19].

Roughly one in every two women in South Asia suffered from anaemia in 2016 — much higher than the global average of one in three women, with the proportion of anaemic women higher in the lowest wealth quintiles (58 per cent), in rural areas (54 per cent) and amongst the Scheduled Castes (56 per cent) and Tribes (60 per cent) [16]. The same trend holds for men, although at much lower proportions than for women (32 per cent in the lowest wealth quintile and 25 per cent in rural areas) [16]. The overlap of violence and poor nutrition, in the above case anaemia, has implications for birth weight, stunting and wasting of children under 5 years of age. Amongst the Scheduled Tribes and Castes, more than two in five children continue to be stunted and one in five are wasted and of low birth weight, versus one in five stunted children amongst the higher wealth quintiles and general classes. Nutritional surveys among tribal groups in India between 1985 and 2008 point to a secular decline in dietary diversity[20], reflecting persistent structural violence against these communities,[21]. In many of these contexts, the cultivated staple grain used to be supplemented by greens, roots, tubers, fruits and fish, secured from rivers and ponds, forests and common lands, ensuring dietary diversity. Such collection tasks are now severely constrained by women's time poverty[22,23], intensified by the lack of infrastructure and services, and also by growing restrictions on these 'common' spaces with the rapid privatization of nature[24,25]. Despite some improvements over the past decade in India, the nutrition situation nevertheless remains serious in rural areas, especially amongst communities who are dependent on agriculture and food production for their livelihoods.

Resilience of food systems is deeply gendered: Why are rural women and men, especially those dependent on farming, the worst off in terms of nutrition and food security outcomes? Several explanations can be found for this paradox, including the lack of adequate attention to agricultural growth and development in states and districts shown to have a high nutrition burden through appropriate investments in research, agricultural extension services, skill building and infrastructure, particularly in ensuring water and land security for the majority of small and marginal holders[26]. The resilience of food systems depends on political, economic and social relations, and these are all deeply gendered. The most marginal groups in India, the Scheduled Tribes and Scheduled Castes, are caught in cycles of indebtedness, increase in severity due to both climate variability and market unpredictability — a phenomenon described by Watts[27] as silent violence. Coping in such contexts reflects positions of structural (dis)advantage and the presence/absence of support systems.

Mechanization and displacement: Technological interventions, particularly the mechanization of transplanting, weeding and harvesting (all female, labour-intensive activities), have led to the displacement of female hired labour belonging to the landless Scheduled Castes and other marginal farming households, while largely bypassing remote rural areas, especially those occupied by the Scheduled Tribes in India. The impacts of mechanization are clearly differentiated, while some women have lost work and incomes, others are potentially engaging in more unremunerated work as family labour and are increasingly time-poor [28,29].

Welfare measures and dietary diversity: Although the national government has not strengthened the agricultural sector through better prices, assured procurement or investments in climate-proofing[30], several state governments in India have sought to ameliorate rural distress and other related effects of agrarian stagnation through welfare measures. Leaders in this respect are the states of Tamil Nadu and Andhra Pradesh with 55 and 48 welfare schemes, respectively. These include the provision of free rice through the universal public distribution system, mixer grinders and limited units of free electricity to reduce the drudgery of domestic labour, and maternity entitlements in both cash and kind (soap, oil, towels, nail clippers for the baby and iron tonics and nutritional supplements for the mother)[31,32]. Although these measures have clearly contributed to food security, the lack of income in the hands of women agricultural workers has meant restrictions on dietary diversity and, in turn, nutrition security.

Gender segregation, lack of recognition and entitlements: Given the gender segregation of labour markets and the wage gaps therein, it is usually men who migrate out of rural areas in search of work. A majority confront poor living and working conditions in migrant destinations, and often return home sick. Considerable portions of their earnings are spent on health expenses. At the same time, their bodies need time for recovery and they are unable to return to work soon[33]. Women left behind with the responsibility for managing agricultural work on family farms, classified as 'unpaid household helpers' in Indian statistics, receive no income, are usually overworked and have no wage-linked benefits[34]. Apart from agricultural work, both household work and child and elder care responsibilities remain assigned to women within existing gender divisions of labour[35], with the additional task of now caring for their men. Overburdened with productive and reproductive work, alongside the lack of recognition and entitlements as farmers in their own right, has pushed women further into relationships of dependency[12], whether with their husbands, traders, contractors and moneylenders, or state offices responsible for welfare provision. Such dependency contributes to the experience of both direct and indirect forms of violence and harassment, whether it be repeated visits to beg dominant men for cash advances, the granting of sexual favours in return for receiving fair wages[36] or simply carrying additional burdens of work. Agricultural work in the present context is then unable to ensure adequate food, incomes or indeed time for the performance of caring and nurturing tasks, which are central to nutrition security. This has adverse impacts on women's own health and child and household nutrition[37]. What is needed is the creative use of technologies, including digital platforms and mobile telephony, to enable women access to information, skills, resources and inclusion in local governance — an opportunity structure that can enable women and the poor to exercise agency, including representation and voice in key decision-making arenas[38]. There is evidence to suggest that women's participation in local government (panchayats) can contribute to improved food intake alongside wider development benefits[39].

Access to resources and opportunities: In rural contexts, evidence from India and other parts of South Asia reveals that successful engagement with higher-return, non-agricultural or agricultural enterprises results from the ownership of land

and other assets (such as wells for irrigation or farm equipment)[31]. Rural women are less likely to be in a position to control productive resources necessary for agriculture in intensely patriarchal contexts and are more restricted in their access to and control over land, energy, water, pasture, forests, agricultural inputs, credit and insurance services, information, technology and markets[40]. Despite progressive legislation in relation to inheritance of agricultural land in India and credit policies that target women's self-help groups across the subcontinent, without mechanisms for enforcing rights, existing gendered power hierarchies ensure that such laws and policies are not operationalized[6]. In fact, there is a greater policing of women's mobility and marriage options[41]. This is because assets are not just material resources with impacts on well-being, but a source of power, symbolizing gendered meanings and identities. Given the social expectation of men as 'providers', land and money are constructed as 'male' assets, so a woman making claims to them can be seen as deviating from the norm of a 'good woman', justifying a violent backlash[7]. Statistical data measuring gender gaps in landholding, employment and education since the 1970s reveal a slight, but not necessarily significant, change over time. Lack of land titles is clearly a problem for women in terms of accessing credit and other resources, specifically from formal institutions such as banks and agricultural cooperatives. Yet field studies seem to indicate that both men and women view household land as joint property, irrespective of who holds the title. In the context of male migration, men are generally supportive of their wives' claims to land; women perhaps are more cautious, cognizant of the need for community recognition, social legitimacy and support[34,42]. However, if the marriage breaks down due to death or separation, they do demand a share of land — widows in India constitute the largest proportion of women landowners[43]. What seems important here is the introduction of gender-specific incentives and entitlements alongside grass-roots mobilization and collective action to claim these resource rights and benefits[44]. In India, the guaranteed right to 100 days of work under the Mahatma Gandhi National Rural Employment Guarantee Act of 2005, or 25 kg of food grains per month under the National Food Security Act of 2013, and the mobilization around these, have provided women with opportunities for bargaining within households and communities.

Although cereal consumption now seems to be adequate, nutrition security remains a harder goal to attain. Apart from changing aspirations within household consumption patterns, reaching the goal of nutrition security is equally dependent on state provisioning of clean drinking water, sanitation and reliable health services. Sanitation, in the Clean India Campaign (Swachh Bharat Abhiyan), is interpreted as universal access to toilets. Although this is undoubtedly important, insufficient attention has been paid to the disposal of solid and liquid waste, contributing to the growing burden of infectious diseases[45]. Similarly, whereas universal health insurance (Ayushman Bharat) may support curative care, primary health care provisioning (which is key to disease prevention) remains weak. The responsibility for both health and nutrition security is now placed on low-paid and poorly trained women community health and nutrition workers, with little power in their local, rural contexts[5]. Rather than devaluing their work and time, often with negative consequences even for their own health, they need to be adequately supported to perform their roles, both through skill-building and the strengthening of referral and back-up health services.

Towards gender equality in food and nutrition: Lack of food can itself be understood as a form of structural or silent violence, or can provoke violence, and experiences of violence can also restrict women's mobility and ability to work and produce food. Central to understandings of food security, then, is an understanding of gendered power relations. However, despite focusing on availability, access, adequacy and absorption[46], current debates fail to recognize the social and gender dynamics underpinning the achievement of food and nutrition security, intertwined at the household level with emotions of love and fear, cooperation and conflict. The idea of 'singlehood for security'[47] could prevail in instances of severe male domination, yet in seeking a combination of economic security and a lower risk of violence, women in general are not necessarily looking for autonomy, but instead a degree of relational mutuality and reciprocity. What becomes clear is the need to move beyond the binaries of women and men, to look at gender as a relational, and therefore, dynamic construct that shifts in response to changes in the policy context, labour market signals and household- and community-level needs and aspirations. Whereas gender relationships signify unequal power, they also alert us to the exercise of everyday agency, the negotiations involved in survival with dignity[48]. Research needs to focus much more on fine-grained analysis of contextual factors that mediate gender relations and food and nutrition security, be it the time and intensity of agricultural work across seasons and cropping patterns, prevailing labour market and childcare arrangements, household socio-economic status, the nature of male contributions to household survival or the mechanisms for inclusion/exclusion in claiming rights and benefits, including social protection and welfare measures seeking to enhance women's agency and bargaining positions. Such an exploration would provide insights into potential entry points for policy that can contribute to positively altering the contexts that constrain choices available to poor rural women, and men. A gender-just food and nutrition security strategy would then involve recognizing women as farmers and producers and strengthening their entitlements[49], while at the same time enhancing male responsibility for sharing the burdens of both production and reproduction[50]. A partial equilibrium analysis, focusing only on food or nutrition security, or the intensity of agricultural work, will not solve the problem. Such a holistic strategy was proposed in the Draft Women Farmer's Entitlement Bill (2013) in India, which defined women as farmers, and sought equal entitlements to all resources and opportunities available to male farmers — whether land, water, credit or technology — alongside equal representation in decision-making around agricultural policies and programmes. Such legislation, if adopted and enforced, has the potential to break the symbolism of women as property, with access to landed property not conditional on their marital status or social position[7], thereby transforming unequal gender relations across social institutions from the household to the state. The draft bill has yet to become law, however. An important dimension that needs serious policy attention is women's time burdens, especially in relation to unpaid care work, as this contributes to the intergenerational transfer of nutritional disadvantage[19]. Functional infrastructure including clean drinking water and clean energy is necessary, alongside public goods such as basic education and healthcare[51]. Equally important is the need for high-quality childcare to ensure feeding and care of young children when the mother is working on the farm[52,53]. Apart from helping to reduce and manage severe forms of malnutrition, creches can also help to reduce women's dependency on extended family members — especially in-laws, who may engage in violence and other forms of control, including denying women food[10,54]. The lack of attention to these services contributes to a perpetuation of both poor nutritional outcomes and gender inequality in South Asia. Mobilization of women and collective action are critical to ensuring the enforcement of law and the implementation of gender-responsive policies and programmes, especially in the face of newer challenges posed by climate change, urbanization, male migration or changing aspirations in the realms of gender divisions of labour, resource control and decision-making. Although positive outcomes are not guaranteed, such public action will go some way to tackling the problem of women's poorly defined rights and, in turn, the threats to their food, nutrition and wider livelihood security, alongside maintaining the pressure for gender equality and justice [55].

## Acknowledgements

This Perspective draws on several years of work in South Asia, particularly India. I thank the many rural women and men who have shared their lives and thoughts with me over the years. Some of these ideas have been refined by my involvement with the DFID-funded Leveraging Agriculture for Nutrition in South Asia (LANSA) consortium (2013–2018) and the BBSRC-funded Transforming India's Green Revolution towards Sustainable Food Supplies (Tigr2ess: 2018–2021), and I acknowledge their support

# REFERENCES

1. Cornwall, A. and EdwardsJ. Introduction: Beijing+20—where now for gender equality?. IDS Bull. 46, 1–8 (2015).

2. Global Hunger Index (GHI, 2019); https: //www.globalhungerindex.org

3. Smith, L. C. and Haddad, L. Reducing child undernutrition: past drivers and priorities for the post-MDG era. World Dev. 68, 180–204 (2015).

4. Comprehensive National Nutrition Survey (CNNS) National Report (Ministry of Health and Family Welfare, Government of India, UNICEF and Population Council, 2019).

5. Delivered by Women, Led by Men: A Gender and Equity Analysis of the Global Health and Social Workforce Human Resources for Health Observer Series No. 24 (WHO, 2019).

6. Rao, N. Assets, agency and legitimacy: towards a relational understanding of gender equality policy and practice. World Dev. 95, 43–54 (2017).

7. Rao, N. Rights, recognition and rape. Econ. Polit. Weekly 48, 18–20 (2013).

8. Rao, N., Gazdar, H., Chanchani, D. and Ibrahim, M. Women's agricultural work and nutrition in South Asia: from pathways to a cross-disciplinary, grounded analytical framework. Food Policy 82, 50–62 (2019).

9.  Usta, J., Makarem, N. and Habib, R. Economic abuse in Lebanon: experiences and perceptions. Violence Against Wom. 19, 356–375 (2013).

10. Lentz, E. C. Complicating narratives of women's food and nutrition insecurity: domestic violence in rural Bangladesh. World Dev. 104, 271–280 (2018).

11. Rao, N. From abandonment to autonomy: gendered strategies for coping with climate change, Isiolo country, Kenya. Geoforum 102, 27–37 (2019).

12. Bellows, A. C., Lemke, S., Jenderedjian, A. and Scherbaum, V. Violence as an unrecognized barrier to women's realization of their right to adequate food and nutrition: case studies from Georgia and South Africa. Violence Against Wom. 21, 1194–1217 (2015).

13. Chilton, M. M., Rabinowich, J. R. and Woolf, N. H. Very low food security in the USA is linked with exposure to violence. Public Health Nutr. 17, 73–82 (2013).

14. García-Moreno, C., Jansen, H., Ellsberg, M., Heise, L. and Watts, C. WHO Multi-Country Study on Women's Health and Domestic Violence Against Women (WHO, 2005).

15. Rao, N. and Pervez, A. in India Social Development Report 2018: Rising Inequalities in India (eds Haque, T. and Reddy, D. N.) 165–173 (Oxford Univ. Press, 2019).

16. National Family Health Survey (NFHS-4), 2015–16: India (IIPS and ICF, 2017).

17. Bellows, A. C. Exposing violences: using women's human rights theory to reconceptualize food rights. J. Agric. Environ. Ethics 16, 249–279 (2003).

18. Ackerson, L. K. and Subramanian, S. V. Domestic violence and chronic malnutrition among women and children in India. Am. J. Epidemiol. 167, 1188–1196 (2008).

19. Chilton, M. M., Knowles, M. and Bloom, S. L. The intergenerational circumstances of household food insecurity and adversity. J. Hunger Environ. Nutr. 12, 269–297 (2017).

20. National Nutrition Monitoring Bureau Diet and Nutritional Status of Tribal Population and Prevalence of Hypertension Amongst Adults: Report on Second Repeat Survey NNMB Technical Report No. 25 (ICMR, 2009).

21. Menon, P., Mani, S. and Nguyen, P. H. How are India's Districts Doing on Maternal, Infant and Young Child Nutrition? Insights from the National Family Health Survey – 4 POSHAN Date Note No. 1 (International Food Policy Research Institute, 2017).

22. Rao, N. and Raju, S. Gendered time, seasonality and nutrition: insights from two Indian districts. Fem. Econ. https://doi.org/10.1080/13545701.2019.163 2470 (2019).

23. Johnston, D., Stevano, S., Malapit, H. J., Hull, E. and Kadiyala, S. Time use as an explanation for the agri-nutrition disconnect: evidence from rural areas in low and middle-income countries. Food Policy 76, 8–18 (2018).

24. Harris, L. M. Gender and emergent water governance: comparative overview of neoliberalized natures and gender dimensions of privatization, devolution and marketization. Gender Place Cult. 16, 387–408 (2009).

25. Mitra, A. and Rao, N. Contract farming, ecological change and the transformations of reciprocal gendered social relations in Eastern India. J. Peasant Stud. https://doi.org/10.1080/03066150.2019.1683000 (2019).

26. NSSO Household Ownership and Operational Holdings in India: NSS 70th Round (Ministry of Statistics and Program Implementation, Government of India, 2013)

27. Watts, M. J. Silent Violence: Food, Famine, and Peasantry in Northern Nigeria (Univ. California Press, 1983).

28. Harriss-White, B. Commercialisation, commodification and gender relations in post-harvest systems for rice in South Asia. Econ. Polit. Weekly 40, 2530–2542 (2005).

29. Agarwal, B. in Technology and Rural Women (ed. Ahmed, I.) 67–150 (George Allen and Unwin, 1985).

30. Swaminathan, M. S. National policy for farmers: ten years later. Rev. Agrar. Stud. 6, 133–144 (2016).

31. Djurfeldt, G. et?al. Agrarian change and social mobility in Tamil Nadu. Econ. Polit. Weekly 43, 50–61 (2008).

32. Swaminathan, M. and Prasad, V. in Towards Universalisation of Maternity Entitlements: An Exploratory Case of Study of the Dr.?Muthulakshmi Reddy Maternity Assistance Scheme, Tamil Nadu Ch. 2 (Public Health Research Network, M. S. Swaminathan Research Foundation and Tamil Nadu Forum for Creche and Child Care Services, 2010).

33. Rao, N. and Mitra, A. Migration, representations and social relations: experiences of Jharkhand labour to western Uttar Pradesh. J. Dev. Stud. 49, 846–860 (2013).

34. Rao, N. Male 'providers' and female 'housewives: a gendered co-performance in rural North India. Dev. Change 43, 1025–1048 (2012).

35. Springer, K. W., Hankivsky, O. and Bates, L. M. Gender and health: relational, intersectional, and biosocial approaches. Soc. Sci. Med. 74, 1661–1666 (2012).

36. Shekhar, H. S. The Adivasi Will Not Dance (Speaking Tiger Books, 2015).

37. Kadiyala, S., Harris, J., Headey, D., Yosef, S. and Gillespie, S. Agriculture and nutrition in India: mapping evidence to pathways. Ann. N. Y. Acad. Sci. 1331, 43–56 (2014).

38. Petesch, P., Smulovitz, C. and Walton, M. in Measuring Empowerment: Cross-disciplinary Perspectives (ed. Narayan, D.) 39–68 (World Bank, 2005).

39. Chattopadhyay, R. and Duflo, E. Women as policymakers: evidence from a randomized policy experiment in India. Econometrica 72, 1409–1443 (2004).

40. The State of Food and Agriculture 2010–2011: Women in Agriculture—Closing the Gender Gap for Development (FAO, 2011).

41. Chowdhry, P. in Caste in Question: Identity of Hierarchy? (ed. Gupta, D.) 1–42 (Sage, 2004).

42. Gender and Land Tenure Security: Challenges and Barriers to Women's Entitlement to Land in India (UN Women and RDI, 2011).

43. Agarwal, B. Widows vs daughters or widows as daughters: property, land and economic security in rural India. Mod. Asian Stud. 32, 1–48 (1998).

44. Kelkar, G. Between Protest and Policy: Women Claim their Right to Agricultural Land in Rural China and India Working Paper 10 (United Nations Research Institute for Social Development, 2016).

45. India State-level Disease Burden Initiative Collaborators. The burden of child and maternal malnutrition and trends in its indicators in the states of India: the Global Burden of Disease Study 1990–2017. Lancet Child Adolesc. Health 3, 855–870 (2019).

46. Rome Declaration on World Food Security (FAO, 1996).

47. Jones, S. Singlehood for security: towards a review of the relative economic status of women and children in woman-led households. Soc. Transit. 30, 13–27 (1999).

48. Connell, R. W. Gender and Power: Society, the Person and Sexual Politics (Stanford Univ. Press, 1987).

49. Sen, A. K. Poverty and Famines: An Essay on Entitlement and Deprivation (Oxford Univ. Press, 1981).

50. Elson, D. Recognize, reduce, and redistribute unpaid care work: how to close the gender gap. New Labor Forum 26, 52–61 (2017).

51. Cavatorta, E., Shankar, B. and Flores-Martinez, A. Explaining cross-state disparities in child nutrition in rural India. World Dev. 76, 216–237 (2015).

52. Prasad, V., Sinha, D., Chatterjee, P. and Gope, R. K. Outcomes of children with severe acute malnutrition in a tribal day-care setting. Ind. Pediatr. 55, 134–136 (2018).

53. Gope, R. K. et?al. Effects of participatory learning and action with women's groups, counselling through home visits and crèches on undernutrition among children under three years in eastern India: a quasi-experimental study. BMC Public Health 19, 962 (2019).

54. Coffey, D., Khera, R. and Spears, D. Intergenerational Effects of Women's Status: Evidence from Joint Indian Households (r.i.c.e., 2015).

55. Dreze, J. and Sen, A. Hunger and Public Action (Clarendon, 1989).

# Chapter 6

# Focus on Sustainable Agricultural Intensification and Small-scale Agriculture for Achieving Zero Hunger

**P.V. Vara Prasad**

*Director of Feed the Future Innovation Lab for Collaborative Research
on Sustainable Intensification at Kansas State University*

The sustainable intensification of agricultural systems offers synergistic opportunities for the co-production of agricultural and natural capital outcomes. Efficiency and substitution are steps towards sustainable intensification, but system redesign is essential to deliver optimum outcomes as ecological and economic conditions change. We show global progress towards sustainable intensification by farms and hectares, using seven sustainable intensification sub-types: integrated pest management, conservation agriculture, integrated crop and biodiversity, pasture and forage, trees, irrigation management and small or patch systems. From 47 sustainable intensification initiatives at scale (each > 104 farms or hectares), we estimate 163 million farms (29 per cent of all worldwide) have crossed a redesign threshold, practising forms of sustainable intensification on 453 Mha of agricultural land (9 per cent of worldwide total). Key challenges include investment to integrate more forms of sustainable intensification in farming systems, creating agricultural knowledge economies and establishing policy measures to scale sustainable intensification further.

The past half century has seen substantial increases in global food production. World population has risen 2.5-fold since 1960 and yet per-capita food production has grown by 50 per cent over the same period[1]. At the same time, evidence shows that agriculture is the single largest cause of biodiversity loss, greenhouse gas emissions, consumptive use of freshwater, loading of nutrients into the biosphere (nitrogen and phosphorus) and a major cause of pollution due to pesticides[2]. This is manifested in soil erosion and degradation, pollution of rivers and seas, depletion of aquifers and climate forcing[3]. As a consequence, efforts have advanced to develop production systems that at least reduce the damage footprint per unit produced[4]. This desire for agricultural systems to produce sufficient and nutritious food without environmental harm, and going further to produce positive contributions to natural, social and human capital, has been reflected in calls for a wide range of different types of more sustainable agriculture[5-7]. The dominant paradigm for agricultural development centres on intensification (productivity enhancement) without integrating sustainability. When the environment is considered, the conventional focus is on reducing negative impacts rather than exploring synergies between intensification and sustainability. There is increasing evidence that sustainability frameworks can improve intensity through shifts in the factors of agricultural production; such as shifts from fertilizers to nitrogen-fixing legumes as part of rotations or intercropping, from pesticides to natural enemies and from ploughing to reduced-intensity tillage.

## Sustainable Intensification

Compatibility between sustainability and intensification was hinted at in the 1980s: first used in conjunction with an examination of African agriculture[8]. Intensification had previously become synonymous with types of agriculture that resulted in environmental harm[9]. The combination of the two terms was an attempt to indicate that desirable outcomes, such as more food and better ecosystem services, need not be mutually exclusive. Both could be achieved by making better use of land, water, biodiversity, labour, knowledge and technologies. SI was further proposed in a number of key commissions, its adoption since increasing from about ten papers annually before 2010 to over 100 per year by 2015[10]. SI is now central to both the UN Sustainable Development Goals and wider efforts to improve global food and nutritional security[11].

SI is defined as an agricultural process or system where valued outcomes are maintained or increased while at least maintaining and progressing to substantial enhancement of environmental outcomes. It incorporates the principles of doing this without the cultivation of more land (and thus loss of non-farmed habitats), in which increases in overall system performance incur no net environmental cost[12-15]. The concept is open, emphasizing outcomes rather than means, applying to any size of enterprise and not predetermining technologies, production type or particular design components. SI seeks synergies between agricultural and landscape-wide system components, and can be distinguished from earlier manifestations of intensification because of the explicit emphasis on a wider set of environmental as well as socially progressive outcomes. Central to the concept of SI is an acceptance that there will be no perfect end point due to the multi-objective nature of sustainability. Thus, no designed system is expected to succeed forever, with no package of practices fitting the shifting dynamics of every location. SI is a necessary but not sufficient component of transformation in the wider food system. Changes in consumption behaviours (for example, in animal products), as well as reductions in food waste, may make greater contributions to the overall sustainability of food and agriculture systems[7], as well as helping to address the challenge of over-consumption of calorie-dense food, which has become a global threat to health. System level changes will be necessary from production to consumption, and eating better is now a priority for affluent countries. At the farm and landscape level, the need for effective SI is nonetheless urgent. Pressure continues to grow on existing agricultural lands. Environmental degradation reduces the asset base[4,16], expansion of urban and road infrastructure captures agricultural land (in the EU28, agricultural land area fell by 31 Mha over 50 years from 1961; in the USA and Canada, 0.5 Mha are lost annually[17,18]); and climate change and associated extreme weather create new stresses, testing the resilience of the global food system[19]. Attempts to implement SI can result in beneficial outcomes for both agricultural output and natural capital[14,20,21]. The largest increases in food productivity have occurred in less-developed countries, mostly starting from a lower output base. In industrialized countries, systems have tended to see increases in efficiency (lower costs), minimizing harm to ecosystem services and often some reductions in crop and livestock yields[22]. However, the global challenge is significant; planetary boundaries are under threat or have been exceeded, world population will continue to grow from 7.6 billion (2018) to 10 billion by 2050[23], and consumption patterns are converging on those typical in affluent countries for some sections of populations, yet still leaving some 800 million people hungry worldwide. One question centres on scale: can agriculture still provide sufficient nutritious food whilst improving natural capital and not compromising other aspects of well-being; and can this occur at a scale to benefit millions of lives, reverse biodiversity loss and environmental contamination and limit greenhouse gas emissions?

A further question centres on how much wider food system changes towards healthier diets could shape the requirements for agricultural production to focus on both food and environmental outcomes: healthier diets tend to be higher in fruit, pulse and nut content, therefore more dependent on pollination services[24]. Healthier diets could also generate enhanced consumer demand for lower pesticide residues. As SI is an umbrella term that includes a wide range of different agricultural practices and technologies, the precise extent of existing SI practice has been largely unknown. We use an analytical framework developed for this global assessment data sets of large-scale changes (by numbers of farms and hectares) that have been made towards SI since 2000. Beyond improved efficiency and substitution to redesign A previous study[25] proposed three non-linear stages in transitions towards sustainability: (i) efficiency, (ii) substitution and (iii) redesign. Although both efficiency and substitution are valuable stages towards system sustainability, they are not sufficient for maximizing co-production of both favourable agricultural and environmental outcomes at regional and continental scales[26].

Efficiency focuses on making better use of on-farm and imported resources within existing system configurations. Many agricultural systems are wasteful, permitting natural capital degradation within the farm or the escape of inputs across system boundaries to cause external costs on-farm and beyond. Post-harvest losses reduce food availability: tackling them contributes directly to efficiency gains and amplifies the benefits of yield increases generated by other means. On-farm efficiency gains can arise from targeting and rationalizing inputs of fertilizer (such as through deep fertilizer placement: used by 1 million farmers in Bangladesh on 2 Mha [27], pesticide and water to focus impact, reduce use and cause less damage to natural capital and human health. Such precision farming can incorporate sensors, detailed soil mapping, GPS and drone mapping, scouting for pests, weather and satellite data, information technology, robotics, improved diagnostics and delivery systems to ensure inputs (for example, pesticide, fertilizer, water) are applied at the rate and time to the right place, and only when needed[17,28,29]. Automatic control and satellite navigation of agricultural vehicles and machinery can enhance energy efficiency and limit soil compaction.

Substitution focuses on the replacement of technologies and practices. The development of new crop varieties and livestock breeds deploys substitution to replace less-efficient system components with alternatives, such as plant varieties better at converting nutrients to biomass, tolerating drought and/or increases in salinity, and with resistance to specific pests and diseases. Other forms of substitution include the release of biological control agents to substitute for inputs); the use of RNA-based gene silencing pesticides; water-based architecture replacing the use of soil in hydroponics; and in no-tillage systems new forms of direct seeding and weed management replacing inversion tillage[14].

The third stage is a fundamental prerequisite for SI to achieve impact at scale. Redesign centres on the composition and structure of agro-ecosystems to deliver sustainability across all dimensions to facilitate food, fibre and fuel production at increased rates. Redesign harnesses predation, parasitism, allelopathy, herbivory, nitrogen fixation, pollination, trophic dependencies and other agro-ecological processes to develop components that deliver beneficial services for the production of crops and livestock[30,31]. A prime aim is to influence the impacts of agroecosystem management on externalities (negative and positive), such as greenhouse gas emissions, clean water, carbon sequestration, biodiversity and dispersal of pests, pathogens and weeds. Whereas efficiency and substitution tend to be additive and incremental within current production systems, redesign brings the most transformative changes across systems. Redesign, however, is a social and institutional as well as agricultural challenge[31,32], as there is a need to create and make productive use of human capital in the form of knowledge and capacity to adapt and innovate, and social capital to promote common landscape scale change, such as for positive biodiversity, water quantity and quality, pest management, and soil health outcomes[33,34].

Negative unintended consequences for human, social and economic capital associated with the system must also be identified and mitigated as part of the redesign process. Redesign is critical as ecological, economic, social and political conditions change across whole landscapes. The changing nature of pest, disease and weed threats illustrates the continuing challenge[35]. New pests and diseases can suddenly emerge in different ways: development of resistance to pesticides; secondary pests outbreaks due to pesticide overuse; climate change facilitating new invasions and accidental long-distance organism transfer. Recent appearances include wheat blast (*Magnoporthe oryzae*) in Bangladesh (2016), and fall armyworm (*Spodoptera frugiperda*) in sub-Saharan Africa (2017). The papaya mealybug (*Paracoccus marginatus*) is native to Mexico, but spread to the Caribbean in 1994 then to Pacific islands by 2002, was reported in Indonesia, India and Sri Lanka by 2008, then to West Africa; the preferred host is papaya, but it has now colonized mulberry, cassava, tomato and eggplant. Each geographic spread, each shift of host, requires redesigns of local agricultural systems and rapid responses from research and extension. Such new pests and diseases may also impact crop pollinators, as illustrated by host shifts and the accidental anthropogenic spread of bee parasites (for example, *Varroa* mites) and pathogens (for example, *Nosema ceranae*)[36].

## Redesign Typology and Methods

We analysed transitions towards redesign in agricultural systems worldwide. We reviewed literature on SI, including meta-analyses and practices, to produce a typology of seven system types that we classify as redesign: (i) integrated pest management, (ii) conservation agriculture, (iii) integrated crop and biodiversity, (iv) pasture and forage, (v) trees in agricultural systems, (vi) irrigation water management and (vii) intensive small and patch systems (Table 6.1). The seven system types span both industrialized and less-developed countries and zones from temperate to tropical. Progress towards SI in developing countries is occurring in the context of the pressing need to implement sustainable development goals for poverty reduction, improved livelihoods and better nutrition by building more-productive and sustainable systems of smallholder agriculture. There are some 570 million farms worldwide, 84 per cent of which are landholdings of less than 2 ha (ref. 37). These small farms make up 12 per cent of total agricultural area, yet produce 70 per cent of food in Africa and Asia. Sustainable intensification will have to be effective worldwide and will have to reach larger numbers of farms in less developed countries: 74 per cent of all farms are in Asia (of which 35 per cent are in China and 24 per cent in India), 9 per cent in sub-Saharan Africa, 7 per cent in central Europe and central Asia, 3 per cent in Latin America and the Caribbean and 3 per cent are in the Middle East and north Africa. Owing to the average size of the 4 per cent of farms in industrialized countries, the choices made by a single farmer can have landscape-wide consequences. We have screened 400 SI projects, programmes and initiatives worldwide (drawn from literature or existing data sets[20,21,35] and selected those implemented to a scale greater than 104 farms or hectares. Our intention is not to map all innovation for SI worldwide, but to assess where innovation has scaled to have potentially positive outcomes on ecosystem services as well as agricultural objectives across landscapes.

Our results revealed that there are 47 SI initiatives exceeding the 104 scale, of which 17 exceed the 105 threshold and 14 the 106 scale (Supplementary Table 6.1; Figures 6.1 and 6.2). Many SI initiatives worldwide show promise but remain limited in scale (either demonstrating locally dependent conditioning, or the lack of attention to scalar mechanisms). We estimate from these projects, initiatives in some 100 countries that 163 million farms have crossed an important substitution–redesign

**Table 6.1: Redesign Typology and Examples of Sub-types of Intervention**

| Redesign type | Illustrative redesign sub-types of intervention |
|---|---|
| 1. Integrated pest management | Integrated pest management through farmer field schools |
| | Integrated plant and pest management |
| | Push-pull systems |
| 2. Conservation agriculture | Conservation agriculture practices |
| | Zero- and low-tillage |
| | Soil conservation and soil erosion prevention |
| | Enhancement of soil health |
| 3. Integrated crop and biodiversity redesign | Organic agriculture |
| | Rice-fish systems |
| | Systems of crop and rice intensification |
| | Zero-budget natural farming |
| | Science and technology backyard platforms |
| | Farmer wisdom networks |
| | Landcare and watershed management groups |
| 4. Pasture and forage redesign | Mixed forage-crop systems |
| | Management intensive rotational grazing systems |
| | Agropastoral field schools |
| 5. Trees in agricultural systems | Agroforestry |
| | Joint and collective forest management |
| | Leguminous fertilizer trees and shrubs |
| 6. Irrigation water management | Water user associations |
| | Participatory irrigation management |
| | Watershed management |
| | Micro-irrigation technologies |
| 7. Intensive small and patch scale systems | Community farms, allotments, backyard gardens, raised beds |
| | Vertical farms |
| | Group purchasing associations and artisanal small producers (in community-supported agriculture operations, tekei groups, guilds) |
| | Micro-credit groups for small-scale intensification |
| | Integrated aquaculture |

This is an illustrative list of sub-types. Some sub-types span a number of types (for example, organic agriculture also appears in elements of 4 and 7). Community supported agriculture operations are group purchasing associations in North America and the UK, tekei groups are in Japan, guilds in France, Belgium and Switzerland.

threshold, and are using SI methods, in at least one farm enterprise, on an area approaching 453 Mha of agricultural land (not counting the SI initiatives in home and urban gardens and on field boundaries). This comprises 29 per cent of all farms worldwide; and 9 per cent of agricultural land (total worldwide crop and pasture land is 4.9 109 hectares). We note that this global assessment might imply numbers

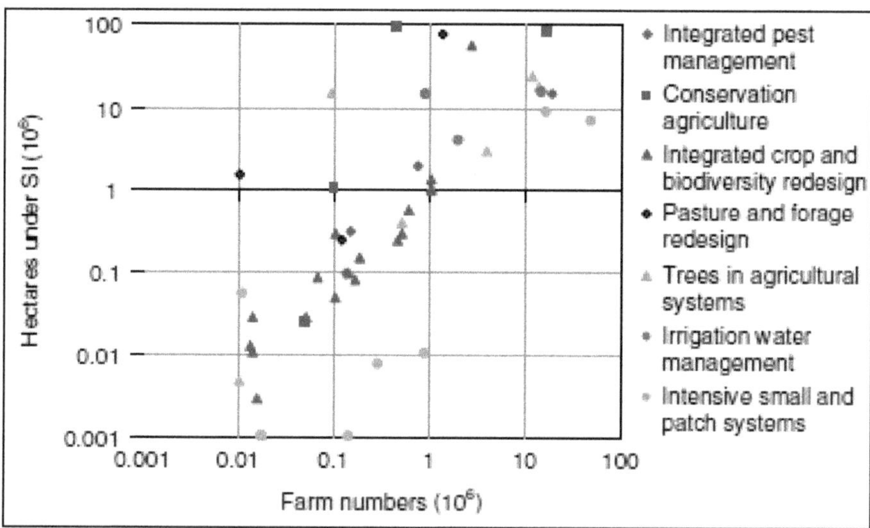


**Figure 6.1: Farm Numbers and Hectares under Seven Types of Sustainable Intensification (47 Initiatives).**

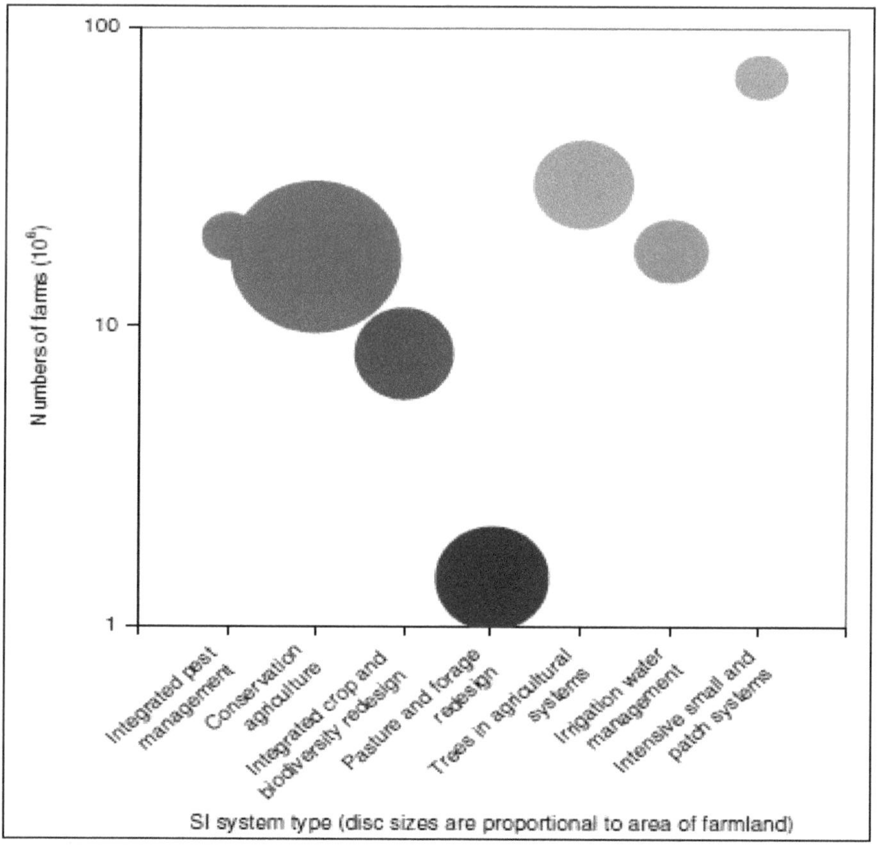

**Figu** **Fig. 2 |** Seven types of sustainable intensification (47 initiatives).

of farms and hectares are fixed: on the ground, there will be a flux in numbers as a result of both adoption and dis-adoption. This may arise from farmer choice and agency, but equally from the actions of vested interests, agricultural input companies, consolidation of small farms into larger operations, changes in agricultural policy or shifts in market demand, and discrepancies between on-paper claims and what farmers have implemented. We have also not included apparent adoption in this assessment: for example, EU regulations require all farms to use integrated pest management, but this has not yet led to significant uptake of agricultural practices that significantly benefit ecosystem services[21].

## The Co-creation of Agricultural Knowledge Economies

For SI to have a transformative impact on whole landscapes, it requires cooperation, or at least individual actions that collectively result in additive or synergistic benefits. For farmers to be able to adapt their agro-ecosystems in the face of stresses, they will need to have the confidence to innovate. As ecological, climatic and economic conditions change, and as knowledge evolves, so must the capacity of farmers and communities to allow them to drive transitions through processes of collective social learning. This suggests a valued property of intrinsic adaptability, whereby interventions that can be adapted by users to evolve with changing environmental, economic and social conditions are likely to be more sustainable than those requiring a rigid set of conditions to function. Every example of successful redesign for SI at scale has involved the prior building of social capital[32], in which emphasis is paid to:

1. Relations of trust,
2. Reciprocity and exchange,
3. Common rules, norms and sanctions and
4. Connectedness in groups.

As social capital lowers the costs of working together, it facilitates co-operation, and people have the confidence to invest in collective activities, knowing that others will do so too. They are also less likely to engage in free-rider actions that result in resource degradation. This suggests the need for new knowledge economies for agriculture[38]. The technologies and practices increasingly exist to provide both positive food and ecosystem outcomes: new knowledge needs to be co-created and deployed in an interconnected fashion, with an emphasis on ecological as well as technological innovation. This includes the need to rebuild extension systems and extend them to environmental as well as agronomic skills, with farmer field schools already dense enough in some locations that they have transformed knowledge co-creation and behavioural change[34]. Important examples in industrialized countries include the Landcare movement in Australia with 6,000 groups, farmer-led watershed councils and the Long-term Agro-ecosystem Research Network in the USA, the French network of agroecology farms and the 49 Farmer Cluster Initiatives in the UK[39,40]. These have created platforms for creation of practices to address locally specific problems of erosion, nutrient loss, pathogen escape and waterlogging. In Cuba, the *Campesinoa- Campesino* movement integrates agroecology into redesign, with knowledge and technologies spread through exchange and cooperatives: productivity of 100,000 farmers increased by 150 per cent over ten years and pesticide use fell to 15 per cent of former levels[41]. In West Africa, innovation platforms have increased yield in maize and cassava systems[42], and in Bangladesh have resulted in the development and spread of direct seeded and early-maturing rice[43]. In China, Science and Technology Backyard platforms operate in 21 provinces covering many crops: wheat, maize, rice, soybean, potato, mango and lychee[44]. Science and Technology Backyard platforms bring agricultural scientists to live in villages, and use field demonstrations and farm schools to engage farmers in developing innovations: reasons for success centre on in-person communication, socio-cultural bonding and the trust developed among farmer groups of 30-40 individuals.

## Next Steps to a Tipping Point

This analysis shows that the expansion of SI has begun to occur at scale across a wide range of agro-ecosystems. The benefits of both scientific and farmer input into technologies and practices that combine crops and animals with appropriate agro-ecological and agronomic management are increasingly evident. The associated creation of novel social infrastructure results in both flows of information and builds trust among individuals and agencies. This should result in the improvement of farmer knowledge and capacity through the use of platforms for cooperation together with digital communication technologies. The key question thus centres on what could happen next. SI has been shown to increase productivity[4,5,] raise system diversity[3], reduce farmer costs[20,22,30], reduce negative externalities[12,13,30] and improve ecosystem services[26,30]. There is thus a range of potential motivations for farmers to adopt SI approaches, and for policy support to be provided by national government, third sector and international organizations. SI requires investment to build natural, social and human capital, so is not costless[6,7]. In all 47 initiatives, there are differences in SI adoption by types of farm, farmers and SI sub-type. All innovations begin on a small scale, yet here expanded to exceed the 104 scale for farm numbers and/or hectares. But several hundred more projects remain small in scale or are at early stages of development. In some cases, innovations started with efficiency or substitution interventions and then spread to redesign[31]. In every case, social capital formation leading to knowledge co-creation has been a critical pre-requisite. In every case, too, farmer benefit (for example, food output, income, health) will have been demonstrated and understood. In most contexts, though, state policies for SI remain poorly developed or counter-productive. In the EU, farm subsidies have increasingly been shifting towards targeted environmental outcomes rather than payments for production, a process the UK government has plans to accelerate [45,46,] but this seldom guarantees synergistic benefits across whole landscapes. Several countries have offered explicit public policy support to social group formation, such as for Landcare (Australia), watershed management (India), joint forest management (India, Nepal, Democratic Republic of Congo), irrigation user groups (Mexico) and farmer field schools (Indonesia, Burkina Faso). In India's state of Andhra Pradesh, the state government has made explicit its support to zero-budget natural farming (local form of uncertified organic farming), aiming to reach 6 million farmers by 2027[47]; in Bhutan and the Indian states of Kerala and Sikkim, policy commitments have been made to convert all land to organic agriculture; the greening of the Sahel through agroforesty began when national tree ownership regulations were changed to favour local people[12].

Innovation, coordination, greening and sharing as key parts of a new strategy for SI[48]. At the same time, consumers are increasingly playing a role in connecting directly with farmers in affluent countries, such as through group purchasing schemes, farmers' markets and certification schemes, which may in turn change consumption choices[49].

With this growing understanding of the positive roles governments can play in structuring incentives and policies, as well as supporting agricultural knowledge economies, we anticipate that SI may be at a tipping point[24]. A further small increase in the number of farms successfully operating re-designed agricultural systems could lead rapidly to re-design of agriculture on a global scale. To transform agriculture to provide comprehensive sustainably intensified systems that can deliver adequate, healthy food for all people, will require the integration of different redesign types to create system-wide transitions, and the internalization of agricultural externalities into prices or through consumer demand. Our hypothesis is that important synergies are occurring, where redesigned systems will deliver more than the sum of the parts, and that when more than one SI sub-type is combined, the likelihood will increase that redesigned systems will be better fitted to local circumstances and thus be more resilient. In the 47 initiatives analysed here, we scored for the number of types used in each initiative (Table 6.2).

**Table 6.2: Number of Redesign Types of SI Deployed in Each of 47 Initiatives, by Farm and Hectare Number and Proportions**

| | Number of redesign types deployed | | | | |
| --- | --- | --- | --- | --- | --- |
| | 1 | 2 | 3 | 4 | 5-7 |
| Farms (millions) | 50.7 | 132.5 | 16.1 | 1.0 | 0.0 |
| Proportion of farms in each redesign type | 25.3% | 66.1% | 8.0% | 0.5% | 0.0% |
| Area (Mha) | 170.2 | 240.5 | 32.8 | 19.5 | 0.0 |
| Proportion of hectares in each redesign type | 36.8% | 51.9% | 7.1% | 4.2% | 0.0% |

Most initiatives are deploying one (25 per cent of farms, 37 per cent of hectares) or two (66 per cent of farms, 52 per cent of hectares) types. The most-common paired combinations were integrated crop and biodiversity redesign with either integrated pest management, conservation agriculture and soil health, agroforestry and irrigation management. The most-common deployment of only one sub-type was trees in agricultural systems. This suggests a clear challenge centred on further integration: this might include, for example, combining conservation agriculture for soil health with integrated watershed management, nutrient recycling and integrated pest management. There is much to be done to ensure agricultural and food systems worldwide increase the production of nutritious food while ensuring positive impacts on natural and social capital. Some efficiency based initiatives are reaching large numbers of farmers, such as the 21 million reducing fertilizer use in China[50].

We conclude that a transition from efficiency through substitution to redesign will be essential, suggesting that the concept and practice of SI in agriculture will be a process of adaptation, driven by a wide range of actors cooperating in new agricultural knowledge economies. This will still need farmers and society to invest in SI, not just for the sake of sustainability, but for livelihoods and profitability. There are risks like technologies could be dis-adopted, advances lost, and competing interests could co-opt and dilute innovations. Positive changes towards consuming healthier food and reductions in food waste may also not occur, putting more pressure on farmers to produce more food at any cost.

It is recommended that three key questions will need addressing for SI to fulfil its potential across agro-ecosystems worldwide:

1. What further evidence is needed to spread SI innovations as options of choice and best practice globally, thus contributing to further progress towards global food security and landscape- wide benefits for natural capital?

2. How can agricultural systems be redesigned to ensure it is more profitable to maintain, rather than erode, natural capital?

3. How can national policy support for the mainstreaming of SI be strengthened and implemented within and across all countries?

# REFERENCES


1. FAOSTAT database (Food and Agriculture Organization, 2017).

2. West, P. C. *et al.*, Leverage points for improving global food security and the environment. *Science* **345**, 325–328 (2014).

3. Pywell, R. F. *et al.*, Wildlife friendly farming increases crop yield: evidence for ecological intensification. *Proc. R. Soc. Lond. B* **282**, 20151740 (2015).

4. Rockström, J. *et al.*, Sustainable intensification of agriculture for human prosperity and global sustainability. *Ambio* **46**, 4–17 (2017).

5. *The Future of Global Food and Farming* (Foresight, Government Office for Science, 2011).

6. *Save and Grow: Maize, Rice and Wheat – A Guide to Sustainable Crop Production* (FAO, 2016).

7. Benton T. G. in *Routledge Handbook of Food and Nutrition Security* (eds Pritchard, B., Ortiz, R. and Shekar, M.) Ch. 6 (Routledge, Abingdon, 2015).

8. Pretty, J. The sustainable intensification of agriculture. *Nat. Resour. Forum* **21**, 247–256 (1997).

9. Collier, W. L. *et al.*, Recent changes in rice harvesting methods. Some serious social implications. *Bull. Indones. Econ. Stud.* **9**, 36–45 (1973).

10. Gunton, R. M. *et al.*, How scalable is sustainable intensification? *Nat. Plants* **2**, 16065 (2016).

11. *Sustainable Development Goals* (UN Sustainable Development Platform, 2017).

12. Godfray, H. C. J. *et al.*, Food security: the challenge of feeding 9 billion people. *Science* **327**, 812–818 (2010).

13. Smith, P. Delivering food security without increasing pressure on land. *Glob. Food Secur.* **2**, 18–23 (2013).

14. Pretty, J. and Bharucha, Z. P. Sustainable intensification in agricultural systems. *Ann. Bot.* **114**, 1571–1596 (2014).

15. Geertsema, W. *et al.*, Actionable knowledge for ecological intensification of agriculture. *Front. Ecol. Environ.* **14**, 209–216 (2016).

16. Hallman, C. A. *et al.*, More than 75 per cent decline over 27 years in total flying insect biomass in protected areas. *PLoS ONE* **12**, e0185809 (2017).

17. Buckwell, A. *et al.*, The Sustainable Intensification of European Agriculture. (RISE Foundation: 2014).

18. Francis, C. A. *et al.*, Farmland conversion to non-agricultural uses in the US and Canada: current impacts and concerns for the future. *Int J. Agric. Sust.* **10**, 8–24 (2012).

19. IPCC *Climate Change 2014: Impacts Adaptation and Vulnerability* (eds Field, C. B. *et al.*) (Cambridge Univ. Press, Cambridge, 2014).

20. Pretty, J. *et al.*, Resource-conserving agriculture increases yields in developing countries. *Environ. Sci. Technol.* **40**, 1114–1119 (2006).

21. Pretty, J. *et al.*, Sustainable intensification in African agriculture. *Intern. J. Agric. Sustain.* **9**, 5–24 (2011).

22. Reganold, J. P. and Wachter, J. M. Organic agriculture in the twenty-first century. *Nat. Plants* **2**, 15221 (2016).

23. *World Population Prospects: 2017 Revision* (UN Department of Economic and Social Affairs, 2017).

24. Smith, M. R. *et al.*, Effects of decreases of animal pollinators on human nutrition and global health: a modelling analysis. *Lancet* **386**, 1964–1972 (2015).

25. Hill, S. Redesigning the food system for sustainability. *Alternatives* **12**, 32–36 (1985).

26. Sandhu, H. *et al.*, Significance and value of non-traded ecosystem services on farmland. *PeerJ* **3**, p.e762 (2015).

27. Mulligan, K. *Fertilizer Deep Placement* (Feed the Future, USAID, Washington DC, 2016).

28. Garbach, K. *et al.*, Examining multi-functionality for crop yield and ecosystem services in five systems of agroecological intensification. *Intern. J. Agric. Sust.* **15**, 11–28 (2017).

29. Lampkin, N. H. *et al.*, *The Role of Agroecology in Sustainable Intensification* (Organic Research Centre, Elm Farm, Game and Wildlife Conservation Trust, 2015).

30. Gurr, G. M. *et al.*, Multi-country evidence that crop diversification promotes ecological intensification of agriculture. *Nat. Plants* **2**, 16014 (2016).

31. Gliessman, S. R. and Rosemeyer, M. (eds) *The Conversion to Sustainable Agriculture: Principles, Processes, and Practices* (CRC, Boca Raton, FL, 2009).

32. Hartley, S. E. *et al.*, Defending the leaf surface: intra- and inter-specific differences in silicon deposition in grasses in response to damage and silicon supply. *Front. Plant Sci.* **6**, 35 (2015).

33. Pretty, J. Social capital and the collective management of resources. *Science* **302**, 1912–1915 (2003).

34. *Farmer Field School Guidance Document* (FAO, 2016).

35. Pretty, J. and Bharucha, Z. P. Integrated pest management for sustainable intensification of agriculture in Asia and Africa. *Insects* **6**, 152–82 (2015).

36. Goulson, D. *et al.*, Bee declines driven by combined stress from parasites, pesticides, and lack of flowers. *Science* **347**, 1255957 (2015).

37. Lowder, S. K. *et al.*, The number, size, and distribution of farms, smallholder farms, and family farms worldwide. *World Dev.* **87**, 16–29 (2016).

38. MacMillan, T. and Benton, T. Engage farmers in research. *Nature* **509**, 25–27 (2014).

39. Spiegal, S. *et al.*, Evaluating strategies for sustainable intensification of US agriculture through the Long-Term Agroecosystem Research network. *Environ. Res. Lett.* **13**, 034031 (2018).

40. Campbell, A., Alexandra, J. and Curtis, D. Reflections on four decades of land restoration in Australia. *Rangeland J.* **39**, 405–416 (2017).

41. Rosset, P. M. *et al.*, The Campesino-to-Campesino agroecology movement of ANAP in Cuba: social process methodology in the construction of sustainable peasant agriculture and food sovereignty. *J. Peasant Stud.* **38**, 161–191 (2011).

42. Jatoe, J. P. D. *et al.*, Does sustainable agricultural growth require a system of innovation? Evidence from Ghana and Burkina Faso. *Int. J. Agric. Sust.* **13**, 104–119 (2015).

43. Malabayabas, A. J. B. *et al.*, Impacts of direct-seeded and early-maturing varieties of rice on mitigating seasonal hunger for farming communities in northwest Bangladesh. *Intern. J. Agric. Sust.* **12**, 459–470 (2014).

44. Zhang, W. *et al.*, Closing yield gaps in China by empowering smallholder farmers. *Nature* **537**, 671–674 (2016).

45. *A Green Future: Our 25 Year Plan to Improve the Environment* (Defra, 2018).

46. Morris, C. *et al.*, Sustainable intensification: the view from the farm. *Asp. Appl. Biol.* **136**, 19–26 (2017).

47. Kumar, V. T. *Zero-Budget Nature Farming* (Department of Agriculture, Government of Andhra Pradesh, 2017)

48. *CPC and State Council Guide Opinion on Using New Development Concepts to Accelerate Agricultural Modernisation and Realise Moderate Prosperity Society* (Xinhua, 2016); http: //news.xinhuanet.com/fortune/2016-01/27/c_1117916568.htm

49. Allen, J. E. *et al.*, Do community supported agriculture programmes encourage change to food lifestyle behaviours and health outcomes? New evidence from shareholders. *Intern. J. Agric. Sust.* **15**, 70–82 (2017).

50. Cui, Z. *et al.*, Pursuing sustainable productivity with millions of smallholder farmers. *Nature* **555**, 363–366 (2018).

**A.K. Srivastava**

*Former Director and Vice Chancellor,*
*NDRI, Karnal*

## Indian Agriculture Food Production System Moves from "Deficiency" to "Sufficiency"

India has moved from 'Food deficient' and 'Food Import' country in 1947 with frequent droughts and famines to a 'Food Self-sufficient' and 'Food Export' country in 1980. During the initial post-independence period *i.e.* in 1966, we produced a mere of 75 MT and had to import food for feeding our population. But now we take pride that we are capable to feed our population with the National Food Security Act (NFSA) already in place from last five years since its inception in 2013 that conserves the right of citizens of this nation to food with home grown food grains. This has been possible due to various Food Policies adopted by Government of India time to time as per the need of the nation. During the initial years of independence, there was a need to combat the problems associated with lack of food. Thus, high yielding varieties of rice and wheat were introduced together with massive public investment in agricultural research. The agricultural practices have changed from 'Conventional Farming' to 'Technology-led agriculture' and this approach has resulted in Green, White, Blue and Yellow revolutions with an increase in the production of various food commodities.

From 1951 to 2018-19, food grain, horticultural items, milk and fish production has increased respectively, from 51 to 283.4 MT, 40 to 314.7 MT, 17 to 188 MT, 0.75 to 12.6. Meat production has reached to 7.7 MT with 8.5 per cent increase, while egg production has attained an impressive value of 95.2 billion. This rise in the production has led to increase in the per-capita availability of food grains from 140 kg (in 1950s) to more than 160kg (in 2000s), besides reduction in the food prices for consumers. The NFSA provides a legal entitlement, to receive 5 kg of food grains every month at a subsidized rate, to 75 per cent of the rural and 50 per cent of the urban population. It also has provisions for providing nutritional support to pregnant and lactating women and children in age group of 6 months to fourteen years. This appreciable rise is due to interaction of technology, innovations, services, public policies, farmer's enthusiasm and political will, to name a few. In fact, Indian farmers achieved as much progress in wheat production in four years (1964-68), as was achieved during the preceding 4000 years. The prime mover for farmer's enthusiasm was assured and remunerative market. India has continued to feed its people since the last 5 decades. We are not far behind in achieving the next higher objective of providing nutritional security to all the people of this nation with various schemes of fortification of different food commodities already in place.

Our country is endowed with rich biodiversity in flora as well fauna and accounts for 17 per cent animal, 12 per cent plants and 10 per cent fish genetic resources of the globe. India is a home to more than 17 per cent of world's human and 12 per cent of livestock population (16 per cent of cattle, 57 per cent of buffalo, 17 per cent of goats and 5 per cent of sheep population) with a meagre 2.3 per cent of world's land and 4.2 per cent of the global water. In India, 52 per cent of total land is arable as against 11 per cent in the world and has all 15 major climates of the world. Out of the total, 142 ± 2-million-hectare area of India is cultivated, while 60 million hectare is net irrigated land, with 137 per cent cropping intensity. There are 20 agro-climatic zones and out of 60 soil types, 46 are available in the country. Longer sunshine hours and day length perfectly suit for round the year cultivation of crops. However, it is estimated that the land, water and biodiversity will shrink by 30-50 per cent by 2050. Around 52 per cent of the population of India earns livelihood from agriculture.

As far as animal products are concerned, India is the third largest exporter of meat only after Brazil and Australia, but is the largest exporter of buffalo meat with UP being its number 1 exporter which is further expected to grow due to increasing demand of buffalo meat because of its lean nature and set up of state of the art modern abattoir. Out of 7.37 MT, 3.26 is broiler meat, looking at the huge livestock population these figures are quite dismal. India is the third largest producer of eggs in the world. At present, India is among the top most producers as well as consumer of food grains, fruits and vegetables, milk, poultry, spices and herbs, egg and fishes. Apart from main agricultural crops there is significant achievement in production of coarse cereals specially the millets, and plantation crops such as coconut and spices.

## Are we Prepared to Meet the Projected Demand of Food Production?

The projected food grain production in 2020 and 2050 is respectively 297 and 450 million tonnes. With this production data we will be able to meet the demands of food grains of the said period. There has been a tremendous change in the consumption pattern with respect to food expenditure over the period of time. For urban sector, it has been observed that food grain and edible oils consumption

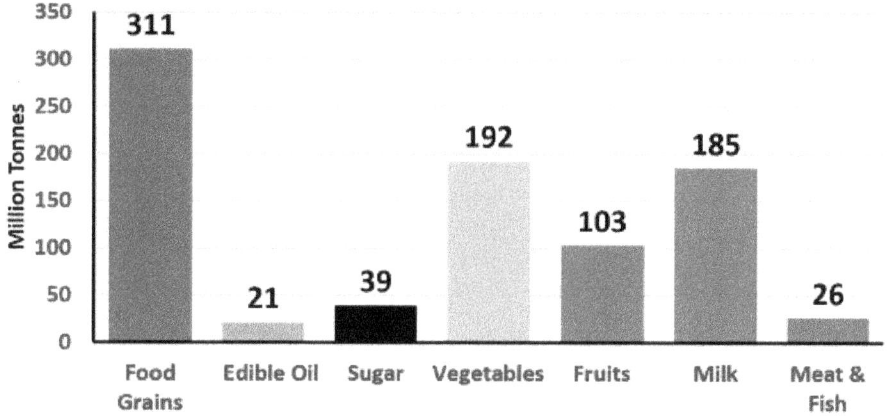


has correspondingly decreased from 31.5 per cent to 24.9 per cent and from 8 per cent to 7 per cent during 1993-94 to 2011-12, while there has been an increase in the consumption of dairy and egg, fish and meat products with values from 17.9 to 20.3 and 6.2 to 7.3 percent, respectively. The consumption of fruits and vegetables has been more or less same (*i.e.* 15 per cent to 14.8 per cent in the said period). Rural sector has also shown a decrease in consumption pattern of food grains (44.8 per cent to 31.7 per cent), while increase in consumption of dairy products (15 per cent to 18.7 per cent), fruits and vegetables (12.2 per cent to 13.8 per cent), edible oils (7 per cent to 7.8 per cent) and egg, fish and meat (5.2 per cent to 7.4 per cent).

Food sector is growing at the rate of 8 per cent per annum. Government is promoting establishment of specialized agro-processing sector through its various schemes. Allocation of MoFPI is doubled from Rs. 715 crore in 2017-18 to Rs. 1400 crore in 2018-19. Large consumer base in our country offers newer opportunities for Indian food processing sector which is still in its nascent stage. India's food processing industry is one of the largest industries in the country - it is ranked fifth in terms of production, consumption, export and expected growth. During the last five years, the growth rate of Indian food processing sector has reached to an impressive figure of 13.7 per cent. However, the Indian food service industry landscape is dominated by unorganized sector, but is experiencing a gradual shift with the expansion of organized sector. The organized food service sector has grown at a CAGR of 10.8 per cent from approximately Rs.750 billion in 2013 to approximately Rs. 1020 billion in 2016. It is expected to capture over 41 per cent of the Indian food service market by 2021. Although, the presence of Indian agricultural produce as well as processed foods in International market is negligible, but country also has competitive edge in terms of raw material, vast pool of skilled manpower and labour. The major trends that are pushing the food service industry towards a high growth trajectory include increasing preference for convenience to eat out, innovations in traditional menu options, growing internet penetration and technological disruptions in food delivery leading to easier and faster availability of food-on-order, *etc.*

In spite of the milestones achieved, Indian food sector faces some major issues: (i) The situation becomes more alarming with more than 80 per cent of the Indian farmers being marginal (*i.e.* upto 1 hectare) and small (1-2 hectares), (ii) Nearly 40 per cent of the marginal and small farmers are most vulnerable, marginalized and food insecure, (iii) The farms are diverse and unorganized, (iv) Sixty per cent of the net cultivated area is rainfed, (v) There has been a tremendous decrease in the arable land/capita (ha) which means there is need to produce more from less (land) for more (population), (vi) With the changing climate the reduction in the Agricultural productivity is estimated to be 25 per cent, (vii) The post-harvest management is the biggest challenge owing to highly inefficient supply chain. The post-harvest loss is estimated to be more than Rs. 80 000 crore annually. The losses in dairy, food grains and horticultural sector being 0.8 per cent, 10 per cent and 30-40 per cent, respectively, (viii) There is a poor emphasis on food safety and quality management systems across the value chain that is one of the major reasons for dismal presence of Indian food products in International market, (ix) There is lack of hygiene in production, storage, processing and marketing of foods. Lastly, the newer emerging threats like changing climate, depleting water table, increasing incidences of pests and insects and decreasing nutritional status of soil must be addressed timely.

## The Status, Problems and Prospects: Indian Food Processing Sector

Food processing operations involve conversion of raw material into wide range of value-added foods that often generates various by-products. The disposal of these bio-wastes (defatted rice bran, defatted corn germ, fruit waste, whey, broken germ, whey, offales, bones, blood, *etc.*) from agricultural activities is a serious problem of India. According to an estimate, agricultural by-products from all crops, livestock and fisheries sector amounts to be about 750 million tonnes annually. At present by-product utilization is approximately 25 per cent and by applying appropriate strategic interventions these can be converted into nutritional food products. There is great potential for conversion of these industrial by-products into value added products for various purposes including industrially important products, ingredient production, in formulation of novel foods *etc.* Indian dairy sector has set an example to be followed by other allied agricultural fields by utilizing whey-the largest by-product of dairy industry both in terms of volume and milk solids. Whey derivatives are becoming popular day by day as quality food ingredients in food and pharma industry. Effective extraction and modification of by-products from agricultural commodities can be implemented as an idea of *'food for food'*, in turn reducing the environmental hazard associated with the bio-waste and also providing job opportunities and healthy food alternatives for the people.

Cold storage facilities are used by all sub-sets of Indian food industry especially for the perishable products. The food service outlets are required to maintain necessary ambient conditions (temperature, humidity, *etc.*). The cold storage market is reportedly growing rapidly at about 20 per cent annual rate since 2011 and was expected to reach to 625 billion by 2017 with the cold storage capacity meeting less than 50 per cent of country's need. The cold chain network is majorly fragmented (3500 companies engaged in cold chain operations in India) with organized players contributing only 8 to 10 per cent of the total cold chain market in India. There is uneven distribution of total cold chain storage capacity with a huge amount of 65 per cent share being contributed by only two states (Uttar Pradesh and West Bengal). The problem gets worsened with more than 35 per cent of India's cold storage facilities having less than 1000MT with more than 75 per cent being dedicated to only potatoes, while 25 per cent committed to other products like fruits, vegetables, pharmaceuticals and processed foods. Approximately, 104 million tonnes of perishable commodities are transported annually, with only 4 million tonnes move through refrigerated mode, while rest via non-refrigerated mode. Out of the total facility available, a major 80 per cent of refrigerated transportation being used by dairy products. Thus, there is a need to establish end-to-end supply chain with modern structures including integrated cold chain solutions, last-mile connectivity, logistics parks, *etc.* Investments in the cold storage industry need to be encouraged through PPP (Public-Private Partnership) and BOT (Build-Operate-Transer) models, with increased incentives and subsidies for the investors.

To conclude, with sustainable high agricultural production, international demand and a strong domestic market, the Indian food industry offers ample scope for large investments in processing technologies, skills and equipment, packaging and storage infrastructure.

## In India Malnutrition is a Bigger Challenge to Address

Malnutrition in infants and young children is one of the most serious health problems in developing world. It influences the normal physical and mental growth and development that resulted in increased mortality and morbidity. It is becoming evident that some of these effects are due to the lack of specific nutrient deficiencies. Across the world about 32 per cent of children below the age of 5 years suffer from being underweight and 39 per cent from stunting. According to the most recent estimate around 854 million people worldwide are undernourished (FAO, 2006). 18.9 per cent of undernourished population and 43 per cent of malnutrition children under 5 are living in India followed by Bangladesh (41 per cent), Afghanistan (31 per cent). Among the developing countries, 44.2 per cent children in India are underweight followed by Vietnam (33.1 per cent) and Pakistan (31.6 per cent). Asia and Africa continents bear the greatest share of all the forms of malnutrition *viz.* stunting, wasting and obesity. As per the National Family Health Survey of India, 55 per cent of children living in rural areas suffer from malnutrition while the corresponding figure for urban areas is 45 per cent. It could be attributed to the lack of diversity and lesser emphasis on complementary feeding. During complementary feeding, most of the developments occur, therefore, introduction of semi-solid foods at the expense of breast milk must provide adequate nutrients for the rapid phase of growth and development. Poor feeding practices as well as lack of suitable complementary foods are responsible for under-nutrition. More than 100 million children under 5 years are underweight and unable to realize their full socio-economic and human potential. This can be seen from the figures corresponding to average consumption of micronutrients. In the urban areas, 56 per cent, 60 per cent and 83 per cent of requirement of RDA of Zinc, Vitamin A and Iron is met while; the corresponding figures for rural areas are 65 per cent, 58 per cent and 90 per cent, respectively. This can be related to the micronutrient deficiency in Indian soils with 48 per cent and 12 per cent deficiency in Zinc and iron, respectively. Obesity prevalence in India has shown considerable increase in the percentile during past two decades. Based on the country income, lower middle-income group has been observed with highest increase in obesity cases with 53 per cent prevalence, followed by low-income group (37 per cent) and high-income group (12 per cent).

### On-going Schemes of Govt. of India: Addressing Mal-nutrition and Hidden Hunger

1. Integrated Child Development Scheme
2. Janani Suraksha Yojana
3. Indira Gandhi Matriya Sahyog Yojana
4. Reproductive Child Health (RCH and RCH-II)
5. National Rural Health Mission (NRHM)
6. Rajiv Gandhi National Crèche Scheme
7. Serva Siksha Abhiyan
8. Mid-Day Meal
9. Kishori Shakti Yojana
10. Women Welfare and Support Programme
11. Old and Infirm Person Annapurna
12. Total Sanitation Campaign (TSC)
13. National Rural Drinking Water Programme
14. Scheme for Empowerment of Adolescent Girls
15. TPDS

Across the world, number of stunted children decreased from 198.4 million to 154.8 million during last two decades while; children suffering from wasting diseases increased from 30.4 million to 40.6 million. Mortality rate of children under 5 years of age in India is reported to be decreased from 114.2/1000 to 61.3/1000 during past 25 years while India is at 12th position with new born mortality rate of 25.4/1000 live births. According to Global Nutrition Report (2016), India has seven indicators under Millennium Development Goals and among these; India has been able to achieve only one indicator (under 5 overweight). In our country, a significant proportion of the population is vulnerable to hidden hunger and very high rates of mortality occur due to coronary heart-diseases (CHDs), cancer and diabetes; all related to diet. Due to lack of diversity in Indian food basket people often deprived off the nutrients from the diet. Presence of anti-nutrients and inhibitors in certain staples further affect the bio-availability of certain key nutrients. Health concerns are attributed to poor nutrition in low-income segments of the population, whereas the well-off strata of the society need to address the health issues primarily due to the changing life-style and food habits.

Iron deficiency is considered to be the commonest worldwide nutritional deficiency and affects approximately 20 per cent of the world population. Women and young children are at risk. According to National Health and Family Health Survey (2016), prevalence of anemia in India is 35.7 per cent in underweight children, 38.4 per cent in stunted children, 21 per cent in wasted children and 58.4 per cent anemic children. This deficiency is partly induced by plant-based diets such as vegetarian diets, containing low levels off poorly bio-available iron. The adverse effects include lower growth rate and impaired cognitive scores in children.

## Malnutrition also Influences the Productive Life of Human Being

Malnutrition is the leading cause of death worldwide in children under the age of five and accounts for deaths of 2.6 million children every year and it leaves millions more life-long impairments. Adequate nutrition during infancy and early childhood is essential to ensure the growth, health and development of children to the full potential. Inappropriate nutrition can also lead to childhood obesity which is an increasing public health problem in our country. The first three years of life provide a critical window of opportunity for ensuring children's appropriate growth and development through optimal feeding. The 1000-day window of time which starts from woman's pregnancy to her child's second year of age is highly crucial.

The gut microbiota are most important factor for human health. Micro-organisms are present throughout the GI-tract, but the major location is large intestine. The gut flora help in degrading food components, production of endogenous mucins and vitamins, prevention of colonization of pathogens and producing essential digestive and protective enzymes. The composition of the gut microbiota in children varies greatly during the first year of life before reaching a more stable and mature composition. The intestinal microbiota of the new born is initially colonized by *Enterobacteria*. In the days after, strict anaerobic bacteria dominate the microbial community. During the first month, *Bifidobacteria* species predominate in the gut, but the introduction of solid foods at around 4–6 months is accompanied by an expansion of *Clostridial* species (*Lachnospiracea, Clostridiaceae,* and *Ruminococcaceae*). Members of the *Ruminococcaceae* family continue to increase in abundance in the following months. By 2–3 years of age, the microbiota composition consists of mainly *Bacteroidaceae, Lachnospiraceae,* and *Ruminococcaceae,* which then remains stable into adulthood. At the later stages of life the microbiota composition becomes less diverse, with a higher *Bacteroides* to *Firmicutes* ratio, an increase in *Proteobacteria* and decrease in *Bifidobacterium.*

The factors that may alter infant gut microbiota are maternal health and microbiota composition, birth mode, antibiotic exposure, breast milk feeding, probiotic and prebiotic feeding *etc.* Recent discoveries suggest that maternal microbes colonize the baby's gut before delivery. At birth, common environment exposures include the maternal vaginal, faecal or skin microbiota. Breastfeeding has been shown to be milk-borne microbiota transplantation. Abnormal breastfeeding and weaning are usually considered as risk factors in the onset of malnutrition. Poverty, malnutrition and infectious disease are interlinked. Poverty is associated with insufficient food and water intake, leading to malnutrition. Malnutrition is associated with a significant impairment of cell-mediated immunity, phagocyte function, immunoglobulin concentration and cytokine production. This situation leads to a high risk of infection like diarrhoea, hence aggravating the nutritional status. According to UN's food and agriculture organization, about 200 million undernourished people living in India. The possible reasons for infant malnutrition are: (i) Decreasing Breastfeeding (deprival of critical immune and growth promoting bioactives), (ii) Lesser emphasis on weaning (nutritionally inadequate and inappropriate), (iii) Declining share of plant-based products in our food basket (aggravated the problem of malnutrition and life-style associated diseases), (iv) Consumption of highly processed foods, frequent use of food additives and presence of processing mediated anti-nutrients

## Strategies to Reduce Infant Malnutrition

WHO and UNICEF's global recommendation for optimal infant feeding as recommended *as Global Strategy* are:

1. **Exclusive breastfeeding for 6 months** (180 days): Even after introduction of complementary foods, breast feeding remains a critical source of nutrients for young infant and child. It provides about one half of an infant's energy needs upto the age of one year, and upto one third during the second year. Studies found that the risk of death from diarrhoea of partially breast-fed infants of 0-6 months of age is 8.6 times the risk for exclusively breast-fed children.

2. **Mother's Milk: Nature's Perfect Food:** Breast milk or mother's milk is probably the first and most diverse kind of functional food which a new born consumes. It is designed by nature to provide all essential nutrients and therapeutic components in desired amount and also in best bio-available form. The bioactive components present in colostrum and mature milk include nutrients, minerals, trace elements and pre-vitamins as well non-nutrients (mostly bioactive) such as immunoglobulin, hormones, growth factors (Insulin-like growth factors), cytokines, prostaglandins, enzymes, lactoferrin, transferrin, nucleotides, polyamines and human milk oligosaccharides (HMO). Breastfeeding continues to offer health benefits into and after toddlerhood. These benefits include; lowered risk of Sudden Infant Death Syndrome (SIDS), faster mental development, lowered incidences of cold and flu, lowered risk of asthma and eczema, decreased risk of obesity later in life, and decreased risk of developing psychological disorder. Breast milk provides a wide variety of proteins that have unique compositional and physico-chemical characteristics that is highly suitable for neonates. In addition to these, they also exhibit several extra-nutritional roles to promote the development and well-being of infants.

3. **Complementary feeding after 6 months:** Complementary foods, commonly known as weaning foods, are semi-solid or solid foods that are used to transition infants from breast milk to an adult diet. In the first

year of life, infants undergo periods of rapid growth when good nutrition is crucial. Energy and nutritional requirements of infant after 6 months of age, exceed what can be supplied by breast milk alone. Complementary foods should be nutritionally adequate, safe and appropriately fed in order to meet the young child's energy and nutrient need. Complementary feeding is often associated with problems, with foods being too dilute, not fed often enough or ion too small amounts or replacing breast milk while being of an inferior quality. Generally cereal components are mainly given as complementary foods, which are inferior in nutrients, has low digestibility and less energy dense. A composite of milk-cereal-fruits and vegetables-pulses could be formulated for complementary feeding employing various processing techniques and may be fortified as per the need of the growing children.

## Role of Milk Nutrients in Well-being

Milk and milk constituents have gained prominence because of increasing scientific evidence pertaining to their health promoting and disease alleviating virtues. Significance of nutraceuticals assume altogether different dimension in our country where rapid rise in malnutrition and incidences of non-communicable diseases is posing newer challenges. It is costing not only 1-2 per cent to National GDP, but adversely affecting the quality human resource as well. Among the functional foods dairy-based products occupy an important place, probably because of the well perceived health benefits associated with consumption of milk and milk nutrients. Milk, dahi and ghee are the three dairy products which has been part of our all religious ceremonies and have been mentioned for their disease preventing abilities in ancient literatures. With changing life-style, there has been increase in the number of chronic diseases at alarming rate. India has attained the first rank in numbers of persons suffering or prone to diabetes, cardiovascular diseases (CVDs) and cancer. Moreover, incidences of infectious diseases are also on rise. One of the common reasons for these diseases could be attributed to impaired or weak immune system. Role of milk nutrients specially the minor milk proteins such as β–Lactoglobulin, α-Lactalbumin and lactoferrin, in modulating the immune system is well documented.

Casein, lactose and milk lipids, the major milk nutrients often serve as base materials for the production of metabolites having positive influence on physiological system and can be termed as nutraceuticals. Richness of milk protein particularly whey proteins, sulphur containing amino-acids like cysteine and methionine assist in enhancing the level of natural antioxidant *i.e.* glutathione. Likewise, abundance of branched chain amino-acids facilitates the effective energy balance during exercises. Serotonin, a biomolecule production is also mediated by milk protein amino-acids. Bioactive lipids mediated compounds including prostaglandins; leukotriene and thromboxane are produced in requisite amounts. Galactose, the hydrolytic product of lactose is essential for the development of vital organs including retina and brain. GMP, a by-product present in cheese whey modulates the bio-synthesis of cholecystokinin, the satiety hormone. Milk phospholipids have attracted the attention of researchers because of their effect on brain health.

Finally, it can be summarized that science led growth and development in food and agriculture will continue to feed future India. But a "Mission Mode" action is required to address the malnutrition. In this regard, livestock, fishery and horticulture will play major role as now the focus should be not on calorie centric but balanced nutritional diet. Let us work together towards the common National goal to provide food and nutritional security for every citizen of the country. The new vision for agriculture food sector for 2050: Let us enhance productivity by 40 per cent ; reduce the hunger and rural poverty by 30 per cent and reduce the emission by 20 per cent.

*(This writeup is part of one of lectures delivered for Dr. D. Sundaresan Memorial Lecture of National Dairy Research Institute, Karnal).*

## Chapter 8

# Production, Food Supply Chains, Trade and Nutrition: The Case of India

Arpita Mukherjee, Angana Parashar Sarma
and Nibha Bharti

*Indian Council for Research in International Economic Relations (ICRIER)*



## Introduction

Agriculture engages over 46 per cent of India's total workforce,[1] and contributes around 17 per cent to the gross domestic product (GDP).[2] India aspires to become a US$5 trillion economy by 2024 and agriculture will be one of the key drivers. Although India's current growth rate had slowed down due to coronavirus (COVID-19) pandemic, the International Monetary Fund (IMF) projected that the growth rate will have a strong rebound, with an estimated growth of over 6 per cent in 2021 (World Economic Outlook, 2020). Agriculture and food processing are among the key sectors, which will help the country to bounce back from current negative to a high growth rate.

India, with its diverse climate and ecology, produces diverse range of crops and is among the top global producers of several agri-food commodities. It is the world's largest producer of milk, pulses, and spices and ranks second in the production of rice, wheat, sugarcane, vegetables, and fruits. It is also one of the

---

1    http://labourbureaunew.gov.in/UserContent/EUS_5th_1.pdf (last accessed August 17, 2020).

2    http://pib.nic.in/newsite/PrintRelease.aspx?relid=186413 (last accessed August 17, 2020).

leading producers of fish and livestock products.[3] Studies have shown that India's agriculture production has been increasing over the years with improved access to various agriculture inputs (such as seeds and fertilisers), better irrigation facilities, and increase in credit coverage [Organisation for Economic Co-operation and Development (OECD), 2018].

Inspite of a large agricultural production, a substantial part of the produce gets wasted in the supply chain due to fragmented supply chain and poor storage and warehousing facilities. While estimates on food loss and wastages vary, there is wastage of an estimated amount of Rs 58,000 crores/year, which is around 7 per cent of India's total food production. This wastage is across the supply chain from production to transportation, processing, retailing and consumption.[4] While government estimates show that around 25 per cent to 30 per cent of fruits and vegetables is wasted due to inadequate logistical facilities, including lack of refrigerated storage, supply chain delays at interstate borders, poor transport and underdeveloped marketing channels, the Food and Agriculture Organisation (FAO) puts this figure at around 40 per cent. Another estimate shows that 21 million metric tonnes of wheat, almost equal to Australia's production, gets wasted and rotten every year due to improper storages. This is an area of concern for a developing country where several people, especially vulnerable groups like women and children suffer from hunger and undernutrition. According to the Global Hunger Index 2019, India ranked 102 out of 117 countries, even behind other South Asian neighbours such as Nepal, Pakistan and Bangladesh. India has the second largest population in the world (1.36 billion) after China (1.43 billion) (United Nation, 2019). It is one of the youngest countries with more than 54 per cent of the total population below the age of 25 years.[5] With a large population, there is a huge pressure on the country to meet the ever-increasing demand for food products. India is both a large importer and exporter of agri-commodities and has a positive trade balance in this sector. India's exports of agriculture food products increased from US$ 28 billion in 2015 to US$ 31 billion in 2019, and its imports decreased from US$ 20 billion in 2015 to US$ 18 billion in 2019.[6] India primarily exports to countries such as the United States and China and imports from countries such as Ukraine, Argentina and Southeast Asian countries such as Indonesia and Malaysia. Some key export items include cereals, fish and crustaceans and meat, while key import items include animal and vegetable fats, and edible fruits, nuts and vegetables. Although India has a negative trade balance in overall goods trade, it has a positive trade balance in agri-food products trade and has strong export potential in this sector.

---

3   <http://www.fao.org/india/fao-in-india/india-at-a-glance/en/#:~:text=India%20is%20the%20world's%20largest,poultry%2C%20livestock%20and%20plantation%20crops.> (last accessed August 17, 2020)

4   https://www.asianage.com/india/all-india/271217/indias-failed-food-system.html (last accessed August 17, 2020)

5   <https://www.mygov.in/newindia/wp-content/uploads/2017/09/newindia301634 1504694260.pdf> (last accessed August 17, 2020)

6   Compiled from World Integrated Trade Statistics (WITS)

Over the past two decades prior to the COVID-19, high growth, rise in income levels, growing population, the increase in demand for a variety of food by the consumers have contributed to the growth of the food processing sector [Confederation of Indian Industry (CII), 2019; RBI, 2019; Ministry of Food Processing Industries, 2017]. Out of the total agricultural output,[7] the share of food processing sector in India was only 46.87 per cent in 2017-18. Comparatively, the share for other developing countries such as Brazil (70 per cent), Malaysia (80 per cent) and Philippines (78 per cent) is much higher [Reserve Bank of India (RBI), 2020]. Although the level of food processing is still low, the food processing sector has been growing at an average annual growth rate of around 5.06 per cent. The sector constituted 8.83 per cent and 10.66 per cent respectively of gross value added (GVA) in manufacturing and agriculture sectors (Ministry of Food Processing and Industries, 2019). This trend is likely to continue with the rapid growth of modern retail, increased urbanisation, changes in the gender composition of work force, improved transportation and changes in consumer perceptions regarding quality and safety. India's market size for food is expected to reach US$ 544 billion by 2020-21 whereas the food industry output is expected to reach US$ 535 billion by 2025-26 (CII, 2019).

With the rise in growth rate, increase in the per capita income of middle and upper-class population and reduction in poverty over the seven decades since India's independence in 1947, there has been a shift from need for 'food security' to need for 'food and nutrition security'. As early as 1993, India adopted the National Nutrition Policy, which documents strong linkages between nutrition and development. The government both at the Centre and state levels have implemented several programmes and schemes towards 'nutrition security for all', which includes National Food Security Act (2013), the National Nutrition Strategy (2017) and the National Nutrition Mission (2017). These polices/initiatives aim to promote convergent approaches that reflect the multidimensional nature of food and nutrition insecurity, and addressing inequalities related to gender, age, disability, income, caste and region.

From the policy perspective, the government's efforts to address malnutrition and food insecurity is linked to reaching its targets under Sustainable Development Goal 2- 'end hunger, achieve food security and improved nutrition and promote sustainable agriculture' and Sustainable Development Goal 3 – 'ensure healthy lives and promote well-being for all at all ages' [Ministry of Statistics and Programme Implementation (MOSPI) and World Food Programme (WFP), 2019].

The Food Safety and Standards Authority of India (FSSAI) was formed in 2011 to ensure food safety and standards along with food security and nutritional safety. It has taken several initiatives such as the 'Eat Right India' movement on July 10, 2018, which is a multi-sectoral effort to nudge citizens to 'eat right/have the right diet'. One of FSSAI's objectives is to reduce the intake of salt, sugar, fat, and phasing-out trans-fats from the diets.

---

7   Total agriculture output includes agricultural crops and livestock sectors.

As per FSSAI, *"Nutrient" means any substance normally consumed as a constituent of food a) which provides energy; or b) which is needed for growth and development and maintenance of healthy life; or c) a deficit of which will cause characteristic bio-chemical or physiological changes to occur. These include vitamins, minerals, trace elements, protein, carbohydrate, fat, fibre and amino acids" — FSSAI, 2012, pp. 2*

Despite the government taking several initiatives to promote nutritional security, studies have pointed out that India suffers from dual problem of malnutrition[8], *i.e.* both undernutrition and overnutrition [Global Nutrition Report, 2020; MOSPI and WFP, 2019; Indian Council for Medical Research and National Institute of Nutrition (NIN), 2019; World Health Organization (WHO), 2018; International Life Science Institute (ILSI) India, 2018].[9] Over the last decade, India has gone through a rapid socio-economic, demographic, lifestyle, nutritional and dietary transitions. However, poverty has still not declined to satisfactory levels and India continues to face the challenges of undernutrition. Even among the relatively affluent urban population, a primary survey study by the author and her research team had found that respondents had a huge deficit in consumption of phyto-nutrients, with intake of fruits and vegetables considerably below the recommended levels of WHO [Mukherjee *et al.*, 2016]. Apart from lack of awareness, increase in income levels, availability of junk food, and factors such as rapid urbanisation has led to challenges of overnutrition, creating risks of non-communicable diseases among all age groups.

Prevalence of undernutrition has been leading to stunted growth and low weight among children, high child mortality, and deficiencies in vitamins and minerals among certain groups of population such as those living under poverty in rural and urban slums and certain socially backward groups such as scheduled castes and scheduled tribes. On the other hand, overnutrition has been a significant challenge for upper and middle-income groups, across different age groups, gender, states and geographical areas, both urban and rural. Studies have shown that due to changes in key demographics and lifestyle of people, and availability of a range of harmful processed food products, there has been an increase in the intake of sugar, salt, and saturated or trans-fats, leading to development of various chronic diseases [WHO, 2018; FSSAI, 2017; Meenakshi, 2016]. The food processing industry in fact, traditionally uses high-levels of sugar, salt and saturated fats in their products, which is a major determinant of the nutrition quality in the processing (or ultra-processing) of foods (FSSAI, 2017). Thus, while food processing can help to reduce food wastage, the type of processed food produced and consumed has an implication on health and that can be driven through right policies. Given this scenario this paper presents *a brief overview of the food production, the food supply chain, trade, nutrition gaps and its impact and policy interventions.*

---

8   Malnutrition refers to deficiencies, excesses, or imbalances in a person's intake of energy and/or nutrients. The term malnutrition addresses 3 broad groups of conditions: Malnutrition in all its forms, includes undernutrition (wasting, stunting, underweight), inadequate vitamins or minerals, overweight, obesity, and resulting diet-related noncommunicable diseases.

9

# Production

India's agricultural food production has increased from Rs. 9.7 trillion in 2009-10 to Rs. 24.93 trillion in 2017-18, accounting for an overall growth rate of around 28.73 per cent.[10] Within sub-categories, milk products, fruits and vegetables, and cereals (which includes rice and wheat) are among the top agricultural produce of India (see Figure 8.1). Among them, output of milk products has grown at a faster rate (66.27 per cent) between 2013-14 and 2017-18 compared to cereals (22.12 per cent).

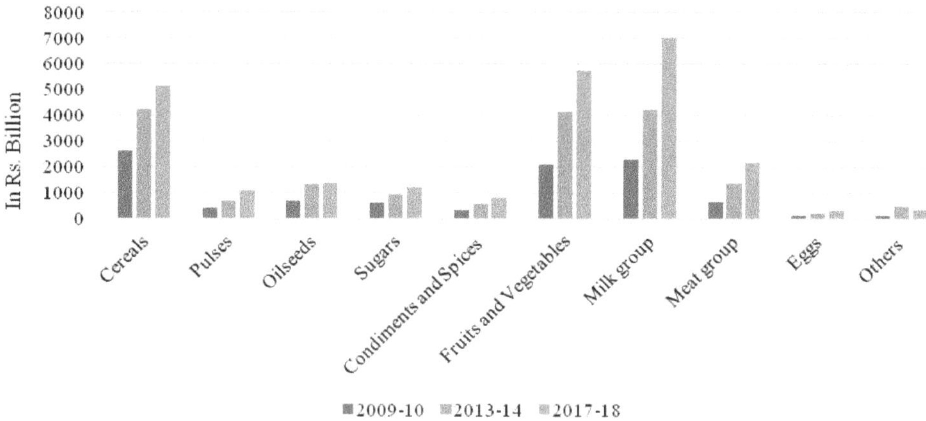


**Figure 8.1: Output of Agricultural Produce by Product Categories.**

*Source*: Compiled from National Account Statistics (NAS), Ministry of Statistics and Programme Implementation (MOSPI). Available at http://www.mospi.gov.in/ publication/ national-accounts-statistics-2019 and http://mospi.nic.in/publication/national-accounts-statistics-2012; (last accessed on July 15 2020).

Specifically, in agricultural food grain production (in volume terms), output has increased over the years from 218.20 million tonnes in 2009-10 to 252.68 million tonnes in 2014-15 to 284.95 million tonnes in 2018-19.[11] However, the share of contribution of the agricultural sector to the total GVA of the country (calculated at constant 2011-12 prices) was 12.81 per cent in 2017-18.[12] The share had decreased continuously from 12.11 per cent in 2011-12 to 9 per cent in 2016-17, before increasing to 12.81 per cent in 2017-18. In case of the food processing sector, the share of contribution to GVA is very low at an average of 1.61 per cent between 2011-12 and

---

10  National Account Statistics, MOSPI

11  Fourth advance estimates of agriculture food grain production, extracted from Annual Reports of Ministry of Agriculture and Family Welfare.

12  The share is less than the share as mentioned in the budget documents, as we have excluded 'forestry sub-sector' from our analysis. As per budget documents and economic survey, government of India categorises agriculture as a broader category of 'agriculture, forestry and fishing' while calculating GVA shares.

2017-18. Out of the total agricultural output,[13] the share of food processing sector in India was at 46.87 per cent in 2017-18 (see Figure 8.2). Between the fiscal years 2008-09 and 2017-18, the sector attained a compound annual growth rate (CAGR) of 12.67 per cent. However, the overall share of food processing in total agricultural output have declined within this time period, after attaining a peak of 53.15 per cent in 2011-12.

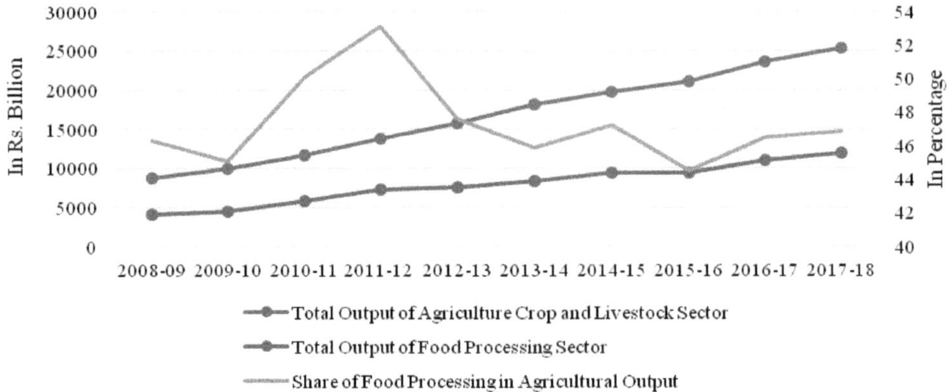


**Figure 8.2: Total Output of Agricultural and Food Processing Sector and Share (2008-09 to 2017-18).**

*Source*: Compiled from National Account Statistics (NAS), National Statistical Survey Office (NSSO)- Industrial Statistics Wing and ASI Summary Results, Ministry of Statistics and Programme Implementation (MOSPI). Available at http://www.mospi.gov.in/publication/national-accounts-statistics-2019 and http://mospi.nic.in/publication/national-accounts-statistics-2012; http://www.csoisw.gov.in/CMS/cms/Feedback.aspx;http://mospi.nic.in/asi-summary -results/107 (last accessed on July 15 2020).

Overall, the food processing sector in India accounts for 32 per cent of the country's total food and grocery market, which is ranked the sixth largest in the world. The sector is ranked the fifth largest industry in the country in terms of production, consumption, exports and expected growth.[14] It is also a major employment provider, contributing to 11.4 per cent of organised manufacturing employment in 2017-18.[15]

## Food Supply Chain

The food supply chain in India is fragmented and constitutes of farmers, commission agents, traders, middlemen, processors, wholesalers, corporate and non-corporate retailers *etc.*, as is given in Figure 8.3. It is important to note that the food supply chain varies by product categories (perishable/non-perishable, conventional

---

13  Total agriculture output includes agricultural crops and livestock sectors.

14  https://www.ibef.org/industry/indian-food-industry.aspx (last accessed on June 18, 2020)

15  Economic Survey 2019-20

vis-a-vis organic) and whether the produce cater to the domestic or export markets. Product traceability is a mandatory requirement for exports of food products like organic food and consumers, especially in developed countries are becoming more aware of the quality and nutrition levels of the food that they consume.

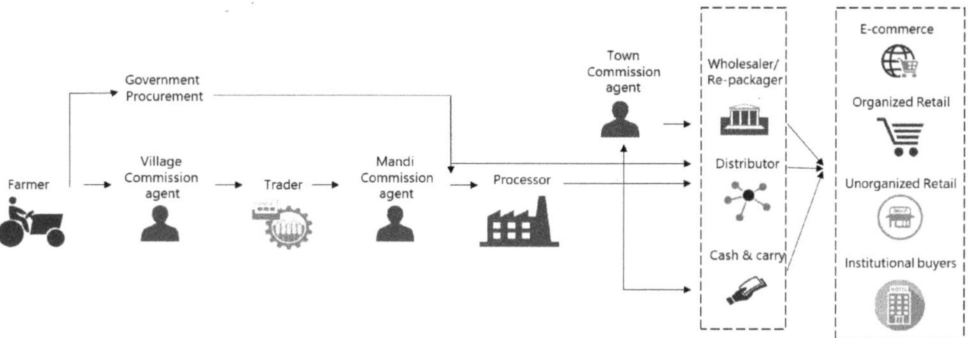


**Figure 8.3: Overview of a Simple Agri-Food Supply Chain in India.**

*Source*: Compiled by Authors.

The key issues leading to food losses and wastage is that the production and consumption hubs are located in different regions across the entire country (for example, see Figures 8.4 (a)–(c) for three key commodities - potato, onion and tomato, which are core to Indian diet). Therefore, there is a need for efficient logistics and supply chain networks, both in terms of physical infrastructure and technology to manage and match the demand and supply.

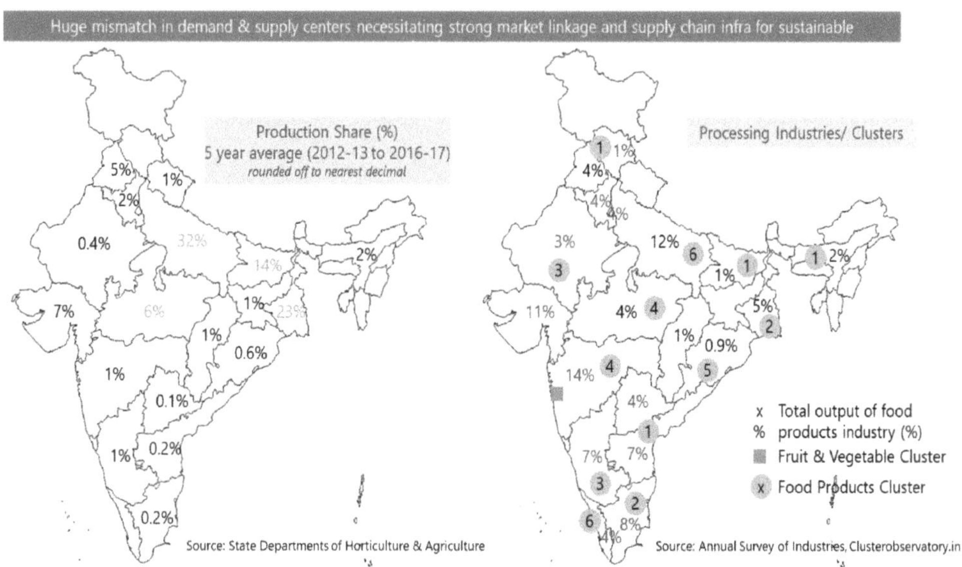


**Figure 8.4(a): Potato: Production Regions and Processing Clusters in India.**

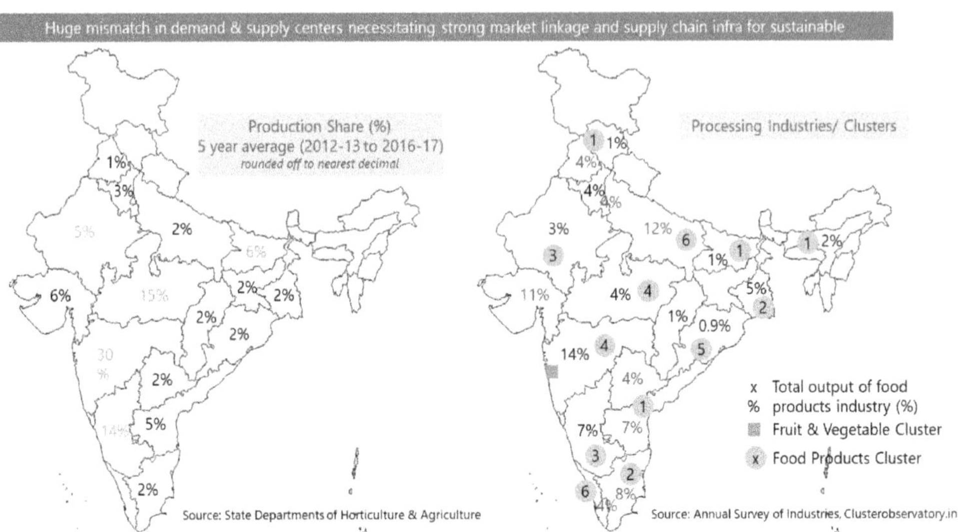


**Figure 8.4(b): Onion: Production Regions and Processing Clusters in India.**

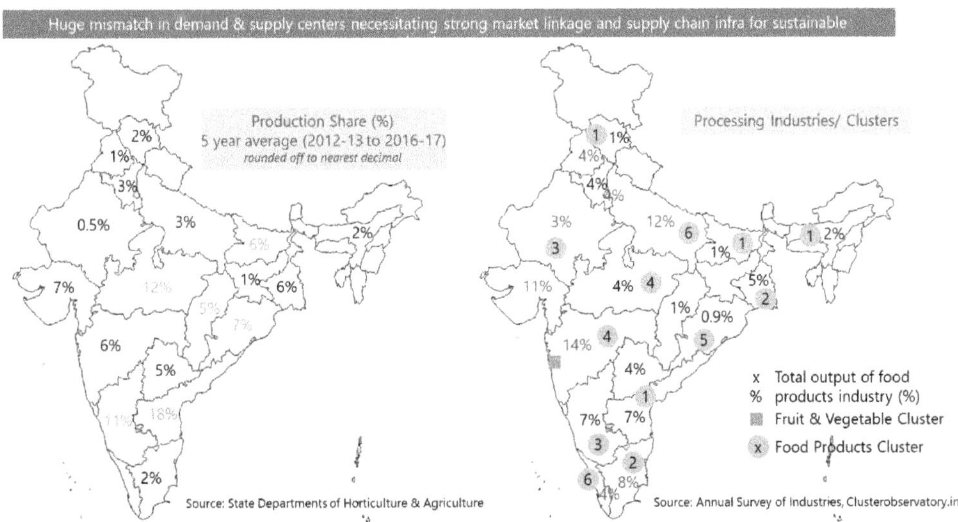


**Figure 8.4(c): Tomato: Production Regions and Processing Clusters in India.**

Studies have shown that farmers have low price realisation and there is huge wastage in the supply chain due to fragmentation, poor storages, inefficient information flow [Negi and Anand, 2016; Pingali *et al.*, 2019, Gokarn and Kuthambalayan, 2017]. Studies also show that India has not been able to reach its full export potential due to fragmented supply chain and gaps in mitigating the demand of importing countries [Mukherjee *et al.*, 2019; Goyal *et al.*, 2017]. Specifically, many of the developed countries want full product traceability from farm to market, and key markets such as the European Union are now focusing on high levels of food safety and standards, which includes lower maximum residue levels than international standards like *Codex* Alimentarius, ensuring that there is no pest infestation, proper food packaging and efficient transportation so that there is no contamination during storage and transport. Many of India's export markets are now focusing on 'nutrition', and for exports, such nutrition related guidelines and requirements (such as ban on trans fats) have to be met.

A survey conducted by the author and her research team shows that in India, unless otherwise put as a requirement by organisations such as the Agricultural and Processed Food Products Export Development Authority (APEDA), sourcing directly from farmers for exports is limited. If the sourcing is from the *mandis*[16], it is difficult to maintain product traceability. There is high incidence of use of chemical fertilizers and pesticides [Goyal *et al.*, 2017], resulting in incidence of diseases such as cancer. Therefore, a huge number of Indian exports are rejected by the importing countries, leading to loss of income of all stakeholders in the export supply chain. Since the country cannot have different food safety standards for imports and domestic production under its commitment to the World Trade Organization (WTO); lower domestic standards for food production and safety would imply lower standards for imports. Therefore, the government is now focusing on the entire food supply chain for efficient production and distribution of nutritious food.

## Trade

India has a positive trade balance in agri-food products and over the years, this gap has increased from US$ 8.06 billion in 2015 to US$ 13 billion in 2019. During the same period, exports increased by 10.40 percent, while imports have decreased by 9.89 per cent after attaining a peak in 2017 at US$24.35 billion (see Figure 8.5).

**Figure 8.5: India Trade in Agriculture Food Products with the World.**

*Source*: Extracted from WITS. Available at https://wits.worldbank.org/WITS/WITS/Restricted/Login.aspx (last accessed on 17 August, 2020)

---

16  Market places for food and agri-commodities, which are mostly a part of the large unorganised markets in India.

In 2019, the top countries for India's agri-food exports and their share were the United States (13.04 per cent), China (7.49 per cent), Iran (6.71 per cent), Vietnam (6.15 per cent) and Saudi Arabia (5.24 per cent); while the top countries that India imported agri-food products from and their share were Indonesia (16.42 per cent), Malaysia (12.90 per cent), Argentina (10.09 per cent), Ukraine (9.07 per cent) and United States (7.52 per cent). In the same year, top exported items included cereals (22.63 per cent), fish and crustaceans, molluscs and other aquatic invertible (19.79 per cent) and meat and edible meat offal (11.05 per cent); while top imported items included animal and vegetable fats (53.93 per cent), edible fruits and nuts (16.87 per cent) and edible vegetables and roots (8.81 per cent).[17]

The issues faced by our exporters include frequent rejection of products for not meeting the MRLs for pesticides, high presence of harmful organisms in the plant produce, pest infestation in fresh food products such as mango and vegetables, lack of harmonisation of standards, not meeting rigid export requirements of advanced countries with respect to technology used or laboratory testing procedures, among others [Mukherjee *et al.*, 2019]. In case of imports, there have been issues related to imports of products of lower standards into the country through different ports of entry and lack of uniformity in compliance and testing facilities across the different ports. However, while this is a key issue for the FSSAI, the WTO's Sanitary and Phytosanitary Agreement (SPS), to which India is a signatory states that imports have to be given equal treatment with domestic production (see Box 1). Hence, it is important to focus on minimum quality of domestic production so that food safety is not compromised.

## Nutrition Security

Over the last decade, with increase in per capita income levels, urbanisation, and changes in people's lifestyles and diets have caused a steady increase in overweight/obesity among people [Dutta *et al.*, 2019; Ford *et al.*, 2017]. On the other hand, challenges of undernutrition continue to a health challenge in India [Nie *et al.*, 2016; United Nations International Children's Education Fund (UNICEF), 2020]. Even after being among the largest food producers in the world, India continues to have significant population who are undernourished.

The increasing prevalence of overweight/obesity accompanying with undernutrition have created the dual problem of malnutrition in India [Global Burden of Disease Study, 2019]. Malnutrition has adverse health implications. For example, studies have shown that over-nutrition leads to chronic diseases such as diabetes, cardiovascular diseases, and overall mortality; while undernutrition is known to be associated with premature mortality and various disabilities in people of all ages (Dutta *et al.*, 2019; Singh, 2020).

The underlying factors behind malnutrition are multi-faceted and differs significantly based on gender, region, age group, and socio-economic factors.

---

17 To calculate trade in agri-food products, the following HS codes at 2-digit level has been used: 02, 03, 04, 07, 08, 09, 10, 11, 12, 15, 16, 17, 18, 19, 20, 21, 22.

**Box 1: WTO Sanitary and Phytosanitary Measures (SPS) Agreement**

The Agreement on the Application of Sanitary and Phytosanitary Measures (known as the "SPS Agreement") entered into force with the establishment of the WTO on 1 January, 1995. The Agreement sets out the basic rules for food safety and animal and plant health standards. It allows countries to set their own standards but also mentions that regulations must be based on scientific analysis. The Agreement states that regulations should be applied only to the extent necessary to protect human, animal or plant life or health and they should not arbitrarily or unjustifiably discriminate between countries where identical or similar conditions prevail. Unjustified discrimination in the use of sanitary and phytosanitary measures are also prohibited in favour of either domestic producers or foreign suppliers.

*Source: Compiled from The WTO Agreement on the Application of Sanitary and Phytosanitary Measures (SPS Agreement), available at https://www.wto.org/english/tratop_e/sps_e/spsagr_e.htm (last accessed August 17, 2020)*

While income levels, sedentary lifestyles and unhealthy eating habits are the key determinants of overweight/obesity (Selvani and Singh, 2018; Bhan *et al.*, 2017); lack of education, low income levels and social disadvantages seem to be the primary reasons behind undernourished population (Chatterjee *et al.*, 2016). Given this scenario, it is extremely important to study malnutrition in India at a deeper level to understand the status of nutritional security in the country.

It is imperative to consider these two aspects of malnutrition as serious health issues and drive appropriate policies around it. Some information on the status of 'undernutrition' and 'overnutrition' is presented below.

### Presence of Undernutrition

Undernutrition[18], has serious health implications as it increases the risk of contracting infectious diseases such as diarrhoea, measles, malaria, pneumonia and can even go on to impact physical and mental development in children if it is severe [Singh, 2020]. In case of India and several other middle-income countries, the inter-generational undernutrition cycle that gets passed on from mothers to children are of even serious nature. Women with undernutrition may go through serious pregnancy-related complications [Nguyen, 2019], undernourished children may not be able to perform well in schools, and undernourished adults in general may be less productive at their workplaces.

In India, the share of adults who are underweight is among the highest in the world [Dutta *et al.*, 2019]. As per the Global Nutrition Report (2020), India is among the 88 countries that are likely to miss global nutrition targets by 2025. The report

---

18  It includes stunting (low height for age), wasting (low weight for height), underweight (low weight for age) and micronutrient deficiencies or insufficiencies (a lack of important vitamins and minerals).

identified that around 37.9 per cent of children under 5 years are stunted and 20.8 per cent are wasted, with stunting being more prevalent in rural areas as compared to urban areas by 10.1 per cent. There is also state-wise variation, with stunting being more prevalent in states such as Uttar Pradesh, Jharkhand, Bihar and Madhya Pradesh as compared to Andhra Pradesh, Karnataka and Haryana [Singh, 2020]. In fact, around half of all the children mortality under 5 years of age is attributable to undernutrition in India.

In the Global Hunger Index rankings of 2019 (which uses four indicators - undernourishment, child wasting, child stunting, and child mortality)[19], India ranked among the bottom 20 (102nd out of 117) low and middle-income countries. Within India itself, there is a wide variation in child mortality rates. In terms of intake of energy and nutrients such as proteins, a study by MOSPI and WFP (2019) pointed out that the current intake level in 2011-12, among the poorest 30 per cent of the income category was lower than the all India average. It is also below the recommended minimum consumption requirement per day of the Indian Council of Medical Research (ICMR) and holds true both in the rural and urban areas.

To reduce the level of undernutrition, the government has come up with several health and awareness programmes and is now focusing on more nutritious grains such as millets. In 2018, Ministry of Women and Child Development, Government of India launched the POSHAN Abhiyan [previously known as the National Nutrition Mission (NNM)], which aims to target problems of malnutrition and its further advanced forms like stunted growth. It has a target of reducing stunting from 38.4 per cent in the year 2017 to 25 per cent by the year 2022 under 'Mission 25 by 2022' (MOSPI and WFP, 2019).

**Presence of Overnutrition**

India is passing through a phase of economic transition with rise in the per capita income among the upper–middle income groups, leading to overnutrition or consumption of unhealthy/wrong-diet. High consumption of unhealthy energy, in turn, increases the risk of a range of diseases such as cancers, cardiovascular diseases, and diabetes [WHO, 2018; Mathur and Pillar, 2019; MOSPI and WFP, 2019]. In addition to factors such as increased globalisation and economic growth, the rising trend in purchases of snacks, hydrogenated edible oils and 'other processed foods' have been some of the key drivers of increased consumption of unhealthy processed food products. Further, it has been driven by the growing Indian policy focus on promoting food processing, as well as liberalisation of FDI in food processing and retailing, which enhanced the availability and affordability of processed foods (Thow *et al.*, 2016).

In recent decades, there has been an increased intake of sugar, oils and highly processed food in Indian diets (for details, see Meenakshi, 2016; Shetty, 2013), with more apparent changes identified in urban India (Gulati and Misra, 2014). Studies have found that the new dietary patterns are contributing towards a number of non-communicable diseases (NCDs) like diabetes, hypertension and cardiovascular

---

19  https://www.globalhungerindex.org/results.html (last accessed August 18, 2020)

diseases in India, which account for a major share of total deaths in the country. For example, overall, cardiovascular diseases contributed to around 28.1 per cent of total deaths in India in 2016 (Prabhakaran *et al.*, 2018). The major constitutes of food that lead to NCDs include saturated and trans fats, sugar and salt.[20] Among these, saturated fats and trans fats found in highly processed foods, bakery products, commercially fried foods and desserts are harmful for health. In fact, trans fats, which are largely present in vanaspati ghee (used for deep fried sweets and snacks), margarine and bakery shortenings, are especially dangerous as they potentially increase the risk of cardiovascular diseases. The FSSAI is engaged in working towards elimination of industrially produced trans fats in the food supply chain by the year 2022, a year ahead of the global target set by the WHO.

The overweight and obesity rates among children have increased in the recent years. Cumulatively, between 2005-06 and 2015-16, it has increased by 19.3 per cent (National Family Health Survey, 2016). State-wise, from 1998 to 2014, the number of states where prevalence of overweight exceeded 20 per cent increased threefold, from two (Kerala and Punjab) to six (Kerala, Punjab, Andhra Pradesh, Uttarakhand, Tamil Nadu, and Karnataka). In some southern states (for example, Tamil Nadu and Kerala) and in wealthier states (for example, Punjab and Himachal Pradesh), the percentage of overweight women was greater than those who were underweight (Meenakshi, 2016).

## Conclusion - Addressing Gaps Through Right Policy

Different policy tools are needed to address the supply chain gaps and dual challenges of under and over nutrition. There is need for targeted interventions through a holistic approach. Globally a mix of policy interventions are used to address issues in food supply chain and nutrition. These include streamlining the food supply chains through infrastructure creation and use of technology, tax regulations (obesity taxes), subsidies for setting up infrastructure, product standard specification and labelling requirements, sourcing strategies for public procurements in schools (mid-day meals), hospitals, *etc.*, and advertisement and marketing restrictions (for example, restriction on advertisements of junk food for children). In addition, governments focus on raising awareness among consumers and processors and strictly implement the food safety and nutrition standards across the supply chain. Policies also focus on creation of supply chain infrastructure and monitoring it to identify and mitigate gaps, if any.

The policy interventions need to focus on the entire supply chain, from the farmers to the consumers. For this, at the farm level, the farmers should be well-equipped with quality inputs and information related to the food safety and hygiene standards, which includes limiting the usage of chemicals and pesticides. This will help India in improving its food quality levels and thus reducing the number of rejections faced in the advanced economies such as the EU. Also, the incidence of diseases related to use of chemicals and pesticides can be addressed.

---

20  https://www.who.int/news-room/fact-sheets/detail/healthy-diet (last accessed July 7, 2020)

The food processors/manufacturers should be guided through policies to reduce wastages such as breakages of rice at the millers/during processing and limit the usage of certain products like trans fats and saturated fats in the food production. The supply chain should be well-integrated from the farm level to the end markets through development of quality infrastructure with respect to storage, warehousing, transportation and logistics services, catering to the requirement of the food products such as perishable and non-perishable. There is a need to develop pan-India cold chain and warehousing grids. In this context, more emphasis should be given on adoption and use of modern technology, data analytics, gaps identification through regular surveys and research and clearly defined short- and long-term mitigation strategies and budgets.

At the consumer level, they must be made aware of the harmful effects of consumption of food with high sugar content, salt, trans fats *etc*. through awareness campaigns at national and sub-national levels. At the same time, consumers should also be made aware of the right diet. These awareness programmes and policies should be marketed well through print, digital and social media to reach a wider consumer base. To encourage consumption of healthy diets, especially among target groups like children, interventions like mid-day meals and the public procurement policies are important. The focus of the government should be to reduce wastages and contamination in the supply chain.

To conclude, it is important for all stakeholders (government, private, international agencies and non-government organisations, academics) to be equally active and work together to achieve the goal of streamlining the supply chain and ensuring nutrition security. Along with the government stakeholders across the entire value chain has to work together to achieve "food and nutrition security".

## REFERENCES

Bhan, N., Millet, C., Subramanian, S.V., Dias, A., Alam, D., Williams, J., and Dhillon P.K. 2017. Socioeconomic Patterning of Chronic Conditions and Behavioral Risk Factors in Rural South Asia: A Multi-Site Cross-Sectional Study. *International Journal of Public Health, 62 (9), pp.1019-1028.* Available at https: //doi. org/10.1007/s00038-017-1019-9 (last accessed August 18, 2020).

Chatterjee, K., Sinha, R.K., Kundu, A.K., Shankar, D., Gope, R., Nair, N., and Tripathy, P.K. 2016. Social Determinants of Inequities in Under-Nutrition (Weight-For-Age) among Under-5 Children: A Cross Sectional Study in Gumla District of Jharkhand, India. *International Journal for Equity in Health, 15, Article Number: 104.* Available at https: //equityhealthj.biomedcentral.com/articles/10.1186/s12939-016-0392-y (last accessed August 18, 2020).

Confederation of Indian Industry. 2019. Indian Food Processing Sector. Trends and Opportunities. *Food Processing Packaging and Technology, Food and Agriculture Center of Excellence.* http: //facecii.in/sites/default/files/food_processing_report_2019.pdf (last accessed on June 17, 2020).

Dorairaj, P., Jeemon, P., Sharma, M., and Roth, G.A. 2018. The Changing Patterns of Cardiovascular Diseases and their Risk Factors in the States of India: The

Global Burden of Disease Study 1990-2016. *The Lancet Global Health*, 6 (12). Available at https: //www.researchgate.net/publication/330157078_The_ changing_patterns_of_cardiovascular_diseases_and_their_risk_factors_ in_the_states_of_India_the_Global_Burden_of_Disease_Study_1990-2016/ link/5c33ee9d92851c22a3638435/download

Dutta, M., Selvamani, Y., Singh, P., and Prashad, L. 2019. The Double Burden of Malnutrition among Adults in India: Evidence from the National Family Health Survey-4 (2015-16). *Epidemiol Health, 41*. Available at https: //pubmed.ncbi. nlm.nih.gov/31962037/

Food Safety and Standards Authority of India (FSSAI). 2017. Report of Expert Group on Consumption of Fat, Sugar and Salt and its Health Effect on Indian Population. Available on https: //www.google.co.in/url?sa=t and rct=j and q= and esrc=s and source=web and cd= and cad=rja and uact=8 and ved=2ahUKEwiSnZeUv5PqAhUSA3IKHdD7DEAQFjAAegQIARAB and url=http per cent 3A per cent 2F per cent 2Fwww.indiaenvironmentportal.org. in per cent 2Fcontent per cent 2F442481 per cent 2Freport-of-the-expert-group- on-issue-of-high-fat-sugar-and-salt-hfss-and-its-health-effects-on-indian- population per cent 2F and usg=AOvVaw0UGq_UMfNLWG5Kr8hzt0fP

Ford, N.D., Patel, S.A., and Narayan, K.M.V. 2017. Obesity in Low- and Middle- Income Countries: Burden, Drivers, and Emerging Challenges. *Annual Review of Public Health, 38, pp. 145-164.* https: //www.annualreviews.org/doi/ full/10.1146/annurev-publhealth-031816-044604

Global Burden of Disease Study. (2019). The Burden of Child and Maternal Malnutrition and Trends in its Indicators in the States of India: The Global Burden of Disease Study 1990–2017. *The LANCET – Child and Adolescent Health, 3 (12), pp. 855-870.* https: //www.thelancet.com/journals/lanchi/article/ PIIS2352-4642(19)30273-1/fulltext#seccestitle260

Global Nutrition Report. 2020. Action on Equity to End Malnutrition. Available at https: //globalnutritionreport.org/reports/2020-global-nutrition-report/

Gokarn, S., and Kuthambalayan, T.S. 2017. Analysis of Challenge Inhibiting the Reduction of Waste in Food Supply Chain. *United States Department of Agriculture.* Available at https: //pubag.nal.usda.gov/catalog/5825414

Goyal, T.M., Mukherjee, A., and Kapoor, A. 2017. India's Export of Food Products: Food and Safety Related Issues and Way Forward. *Indian Council of International Economic Relations (ICRIER) Working Paper, No. 345.* Available at https: //icrier. org/pdf/Working_Paper_345.pdf

Gulati, S., and Misra, A. 2014. Sugar Intake, Obesity, and Diabetes in India. *Nutrients, 6 (12), pp. 5955–5974.* Available at https: //www.ncbi.nlm.nih.gov/pmc/ articles/PMC4277009/

Indian Council for Medical Research (ICMR) and National Institute of Nutrition (NIN) (2019). Current Trends in the Consumption of Fat, Salt and Sugar in India. *International Life Science Institute.* Available at http: //ilsi-india.org/

Current-Trends-in-Food-and-Nutrients-Consumption-in-India/Current-Trends in-Consumption-of-Sugar,-Salt-and-Fat-in-India.pdf

International Life Science Institute-India (ILSI). 2018. Seminar on Recent Developments in Food Science and Technology for Better Nutrition. Available at http: //www.ilsi-india.org/PDF/Report_on_Seminar_on_Recent_ Developments_In_Food_Science_And_Technology_For_Better_Nutrition.pdf

International Monetary Fund (IMF). 2020. World Economic Outlook (WEO). Chapter 1: The Great Lockdown. *International Monetary Fund.* Available at https: // www.imf.org/~/media/Files/Publications/WEO/2020/April/English/text. ashx?la=en ; https: //www.imf.org/~/media/Files/Publications/WEO/2020/ April/English/StatsAppendixA.ashx?la=en

Mathur, P. and Pillai, R. 2019. Overnutrition: Current Scenario and Combat Strategies. *Indian Journal of Medical Research, 149 (6), pp.695-705.* Available at <http: //www.ijmr.org.in/article.asp?issn=0971-5916;year=2019; volume=149;issue=6;spage=695;epage=705;aulast=Mathur>

McMurry, H.S., Shivashankar, R., Mendenhall, E., and Prabhakaran, D. 2017. Insights on Overweight and Obesity. *Economic and Political Weekly, 52 (49).* Available at https: //www.epw.in/journal/2017/49/notes/insights-overweight-and-obesity.html

Meenakshi, J.V. 2016. Trends and Patterns in the Triple Burden of Malnutrition in India. *Agricultural Economics, 47 (S1), pp. 115-134.* Available at https: // onlinelibrary.wiley.com/doi/abs/10.1111/agec.12304

Ministry of Food Processing and Industries. 2017. High-Growth Segments of Indian Food and Beverage Industry. *World Food India 2017. Ernst and Young. Confederation of Indian Industry.* Available at http: //foodprocessingindia.gov. in/publishing/publications/9cfdb26d7526817d8d8bhigh-growth.pdf

Ministry of Food Processing Industries. 2019. Annual Report 2018-19. *Government of India.* Available at https: //mofpi.nic.in/sites/default/files/eng_mofpi_ annual_report_2018-19.pdf

Ministry of Statistics and Programme Implementation (MOSPI) and World Food Programme (WFP). 2019. Food and Nutrition Security Analysis, India, 2019. *Government of India.* Available at http: //www.indiaenvironmentportal.org. in/files/file/Food per cent 20and per cent 20Nutrition per cent 20Security per cent 20Analysis.pdf (last accessed June 8, 2020)

Mukherjee, A., Dutta, S., and Goyal, T. 2016. India's Phytonutrient Report: A Snapshot of Fruits and Vegetables Consumption, Availability and Implications for Phytonutrient Intake. *Academic Foundation.*

Mukherjee, A., Goyal, T.M., Miglani, S., Kapoor, A. 2019. SPS Barriers to India's Agriculture Exports. Learning from the EU Experiences in SPS and Food Safety Standards. ICRIER Report. Available at https: //icrier.org/pdf/SPS_Barriers_ to_India_Agriculture_Export.pdf

Negi, S., and Anand, N. 2017. Factors Leading to Losses and Wastage in the Supply Chain of Fruits and Vegetables Sector in India. *Chapter in Energy Infrastructure and Transportation Challenge and Way* https: //www.researchgate.net/profile/ Saurav_Negi/publication/298891316_Factors_Leading_to_Losses_and_ Wastage_in_the_Supply_Chain_of_Fruits_and_Vegetables_Sector_in_India/ links/56ed3dcc08ae4b8b5e73eea4/Factors-Leading-to-Losses-and-Wastage-in-the-Supply-Chain-of-Fruits-and-Vegetables-Sector-in-India.pdf

Nguyen, H.A. 2019. Undernutrition During Pregnancy. *IntechOpen.* Available at https: //www.intechopen.com/books/complications-of-pregnancy/ undernutrition-during-pregnancy

Nie, P., Rammohan, A., Gwozdz, W., and Sousa, P. 2016. Developments in Undernutrition in Indian Children Under Five: A Decompositional Analysis. *Discussion Paper Series, No. 9893, IZA.* Available at http: //ftp.iza.org/dp9893. pdf

Organisation for Economic Co-operation and Development (OECD). 2018. Review of Agriculture Policies in India. *Trade and Agriculture Directorate Committee for Agriculture.* Available at http: //www.oecd.org/officialdocuments/ publicdisplaydocumentpdf/?cote=TAD/CA(2018)4/FINAL and docLanguage=En

Pingali P.L., Aiyar, A., and Abraham, M. 2019. Transforming Food Systems for Rising India. *Palgrave Studies in Agricultural Economics and Food Policy. Available* at https: //www.researchgate.net/publication/331500711_Transforming_Food_ Systems_for_a_Rising_India

Popkin, B.M. 2014. Nutrition, Agriculture and the Global Food System in Low- and Middle-Income Countries. *Food Policy, 1 (47), pp.91-96.* Available at https: // doi.org/10.1016/j.foodpol.2014.05.001

Reserve Bank of India (RBI). 2020. Food Processing Industry in India: Challenges and Potential. Available at https: //www.rbi.org.in/Scripts/BS_ViewBulletin. aspx?Id=18823

Selvamani, Y., and Singh, P. 2018. Socioeconomic Patterns of Underweight and its Association with Self-Rated Health, Cognition and Quality of Life among Older Adults in India. *PLoS One, 13 (3).* Available at https: //pubmed.ncbi.nlm.nih. gov/29513768/

Shetty, P. 2013. Nutrition Transition and its Health Outcomes. *Indian Journal of Pediatrics, 80 Supplementary 1: S21-S27.* Available at https: //pubmed.ncbi.nlm. nih.gov/23412985/

Singh, A. 2020. Childhood Malnutrition in India. *IntechOpen.* Available at http: // dx.doi.org/10.5772/intechopen.89701

Thow, A.M., Kadiyala, S., Khandelwal, S., Menon, P., Downs, S., and Reddy, K.S. 2016. Toward Food Policy for the Dual Burden of Malnutrition: An Exploratory Policy Space Analysis in India. *Food and Nutrition Bulletin, 37 (3).* Available at https: //journals.sagepub.com/doi/10.1177/0379572116653863

United Nations (UN). 2019. World Population Prospects 2019. *Department of Economic and Social Affairs, Population Division (2019)*. Available at https: //population. un.org/wpp/Publications/Files/WPP2019_ Highlights.pdf

United Nations International Children's Education Fund (UNICEF). 2020. Malnutrition Prevalence remains Alarming: Stunting is Declining too Slowly while Wasting Still Impacts the Lives of far too Many Young Children. Available at https: //data.unicef.org/topic/nutrition/malnutrition/

World Health Organization (WHO). 2018. Obesity and Overweight. Available at: https: //www.who.int/en/news-room/fact-sheets/detail/obesity-and-overweight

Chapter 9

# Overview of Child Health and Nutrition Programme in India

**Sila Deb**

*Addl. Commissioner, Ministry of Health and Family Welfare, GOI*

Ministry of Health and family welfare executes majority of programmes that are aimed to look into the health and nutrition of children. Since India ranks very low in the global list of hunger index and having highest burden of neonatal deaths in the world, the Ministry is framing, planning and running various programmes related to child health and nutrition.

## Status of Child Health in Country

In India some 25 million babies are born annually and about 4.54 million of them are born low birth weight babies that accounts more than 18 per cent of total births. The more serious issue is neonatal deaths. Neonatal deaths (death within one month of birth) are the highest and about 0.57 million deaths of neonates annually accounts for 63.8 per cent of total under-5 deaths and 71.5 per cent of all infant deaths. India has highest burden of neonatal deaths in the globe. India records 0.9 million U-5 deaths annually, of which 0.8 million deaths are infant deaths (deaths within first year of birth) that accounts for 88.8 per cent of all U-5 deaths. As per SRS-2018, 23, 32 and 36 deaths per 1000 account for neonatal, infant and under- 5 mortality respectively.

There is inter-State variation in infant mortality rate, States likes Himachal Pradesh, Utter Pradesh, Odessa, Assam and Chhattisgarh have more than 40 per cent IMR against national average of 32 per cent. At the same time States like Nagaland, Skim, Kerala, Mizoram, Goa and Andaman Nekobar have less than 10 per cent IMR, that stands the target of the country. There is a great variation in IMR within the districts and blocks of a State. Similarly there is disparity in rural IMR that stands 36 per cent and urban IMR that is only 23 per cent.

                                                                    

The reports from National Family Health Survey (NFHS) and Comprehensive National Nutrition Survey (CNNS) indicates improved trends in child under-nutrition and the figures of stunting, wasting and under- weight in U-5 children is given in the Figure 9.1.

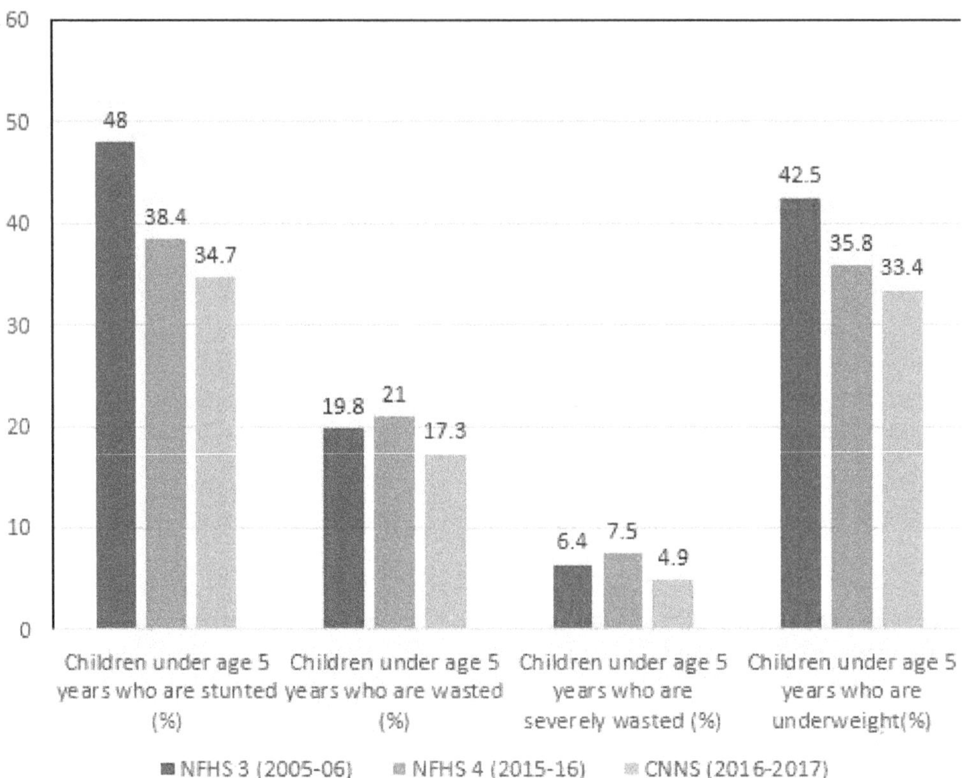


Malnutrition does not refer to only under-nutrition, it also includes over-nutrition also. A person is considered obese when he or she has a body mass index of more than 25 kilograms per square meter. By these standards, at least 20 million women and almost 10 million men in India are now obese. This means India stands fifth when ranked for the most number of obese men in the world and third for the most number of obese women. The data from NFHS indicates that obese women have doubled between 2005 to 2015 from12.6 to 20.6 per cent and obesity in men is not less, it has also doubled during the said period from 9.3 to 18.9 per cent. However, ratio of women with BMI ($18.5kg/m^2$) below normal has reduced from 35.5 per cent to 22.9 per cent and men from 34.2 per cent to 20.2 during the said 10 year period. There has been increase in the childhood obesity during 2005 to 2015 and has increased from 1.5 to 2.1 per cent. This indicates that there has been increase in the intake of calorie rich diets.

Similarly anaemia in U-5 children has reduced from 69.4 per cent to 58.5 per cent, in pregnant women from 57.9 per cent to 50.4 per cent, in women of reproductive age (15-49 years) from 55.3 per cent to 53 per cent and in men (15-49 years) from

24.2 per cent to 22.7 per cent in 10 years from 2005 to 2015 (NFHS-3 and NFHS-4). Prevalence of anaemia in U-5 children and adolescent girls above 40 per cent is a severe public health problem and India is still having it over 50 per cent that need to be reduced.

## Causes of IMR

Preterm birth complications or prematurity is the major cause of IMR and accounts for 30 per cent of all infant mortality. Infections account for 18 per cent, respiratory problems for 13 per cent. Among U-5 children mortality, pneumonia accounts for 17 per cent and diarrhoea for 8 per cent. Under-nutrition is the major issue for the U-5 mortality and contributes to as high as 45 per cent (4.5 Lakh) of U-5 deaths (SRS-2017). The main reason for the under-nutrition is "faulty and suboptimal neonatal and infant feeding practices". Congenital anomalies and vaccine preventable diseases are also very important causes of child mortality. Widespread illiteracy and poverty, lack of adequate public health infrastructure, lack of fresh water and poor sanitation practices are also contributing to the IMR. Deficiency of macronutrients such as proteins, carbohydrates, fats are main causes of under-weight, stunting and wasting whereas deficiency of micronutrients like iron, vitamin B12, folic acid lead to anaemia in children.

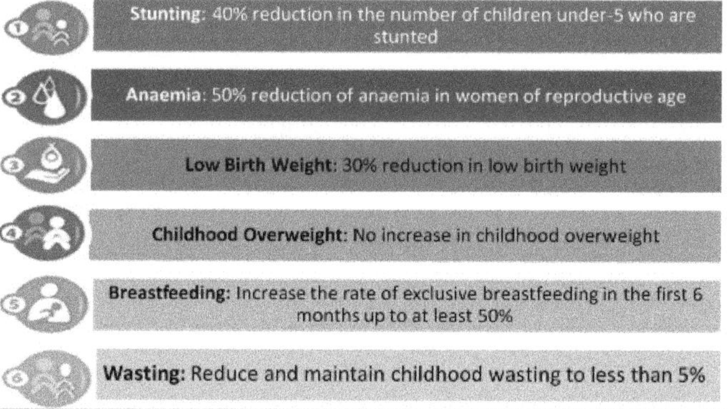


**Sustainable Development Goal 2.2:** By 2030, end all forms of malnutrition, including achieving, by 2025, the internationally agreed targets on stunting and wasting in children under 5 years of age, and address the nutritional needs of adolescent girls, pregnant and lactating women and older persons

## SDG-2.2 Targets

Global targets SDG- 2.2 are set to end all the forms of malnutrition by 2030 to which India is one of the signatories. Internationally it has been agreed to achieve the set targets of malnutrition like stunting, wasting in children of U-5 year age and address the nutritional needs of adolescent, pregnant and lactating women and older persons by 2025.

## Global Nutrition Targets-2025

It is committed to achieve

- ☆ 40 per cent reduction in stunting in U-5 children,
- ☆ 50 per cent reduction in anaemia in women of reproductive age,
- ☆ 30 per cent reduction in low birth weight,
- ☆ No increase in childhood overweight,
- ☆ 50 per cent increase in the exclusive breastfeeding during first 6 months and
- ☆ Reduce and maintain childhood wasting below 5 per cent (fig).

Progress made in India in global targets is detailed in given table.

| Indicator | Global Targets | NFHS-32005-06 | NFHS-4 2015-16 | CNNS 2016-18 | India Nutrition Target 2025 | Progress |
|---|---|---|---|---|---|---|
| Stunting | 40 per cent reduction in the number of children U-5 who are stunted | 48 per cent | 38.4 per cent | 34.7 per cent | 24 per cent | Progressing |
| Wasting | Reduce and maintain childhood wasting to less than 5 per cent | 20 per cent | 21 per cent | 17.3 per cent | < 5 per cent | Off - track |
| Anaemia in WRA | 50 per cent reduction of anaemia in WRA | 55 per cent | 53 per cent | NA | 23 per cent | Off- track |
| Exclusive Breast-feeding | Increase the rate of exclusive breastfeeding in the first 6 months up to atleast 50 per cent | 46 per cent | 54.9 per cent | 58 per cent | 69.2 per cent | On- track |
| Childhood obesity | No increase in childhood overweight | 1.9 per cent | 2 per cent | 2 per cent | 1.9 per cent (maintain) | Off-track |
| Low birth weight | 30 per cent reduction in low birth weight | 22 per cent | 18 per cent | | 30 per cent reduction | Not fixed yet |

## Important Reasons for High Malnutrition in India

The main reasons of malnutrition are chalked out in the figure.

- ☆ More than 50 per cent of adolescent girls, pregnant women and women of reproductive age are anaemic.
- ☆ 51.2 per cent pregnant women observe 4 antenatal checkups, which are very important to reduce intra-uterine malnutrition. These checkups improve the nutritional intake during pregnancy that inturn improves birth weight and mother's health.
- ☆ Early initiation of breastfeeding is very important that alone can reduce 20 per cent neonatal deaths. Breast feeding within one hour of birth is necessary to reduce NMR. However, only 41.6 per cent newborns had early

initiation of breast feeding that is to be increased and is not too difficult because of 80 per cent institutional deliveries in India. Therefore this wide gap is necessary to narrow down through counselling and strict vigilance at the birth centres. Breastfeeding alone can avert 13 per cent U-5 deaths and can help survival of 1.6 lakh babies annually.

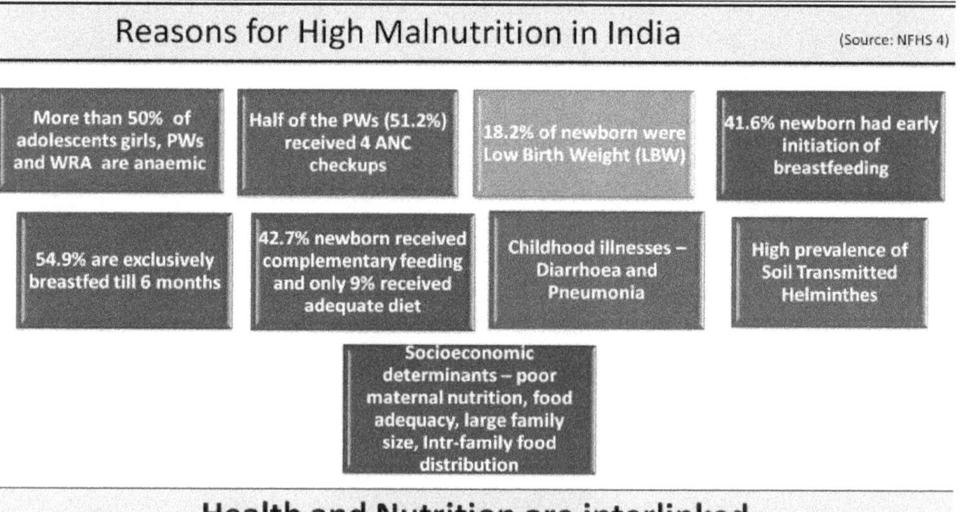


☆ 55 per cent newborns are breastfed exclusively during first 6 months. It is required to enhance it. However, after 6month age only breast feeding is not sufficient and complimentary feeding is necessary.

☆ 43 per cent babies get complementary feeding and only 9 per cent receive adequate diversified diet. To have balanced and nutritious diet, it is necessary to provide diversified diet to the babies that besides grains, should contain fruits, beans and animal products.

☆ The newborns and infants are very prone to infections like pneumonia, diarrhoea and focus should be given to reducing such prevalence through advocacy of proper hygiene, sanitation, safe water utilization and safe and nutritious food.

☆ The children are very much likely to contract soil transmitted helminths like hook worms, round worms. Worm infestation drains the nutrients from the host and the host children become malnourished. Following the proper medical guidelines for deworming of children is an important and effective way to prevent malnutrition in children of U-5 year age.

☆ Socioeconomic determinants are also very important reasons for malnutrition in children. Nutrition of mother plays a vital role in the development of baby in utero and after birth the quality and quantity of mother's milk depends on the nutritional status of lactating mothers. In Indian social setup, the women usually get food for herself after all the members of family are done with. Similarly women get the least out of any special, nutritious dish. Family size plays a great role in the malnutrition of children. More large the family size less access of women to good nutrition. Therefore, it is necessary to change the mindset and focus on better nutrition of pregnant women in the family.

## Implications of Malnutrition

Malnutrition has its implications on various aspects of health, education, economy, productivity and workforce. Malnutrition is the main underlying factor for high morbidity and mortality in U-5 children, neonatal mortality and low birth weight babies. Malnutrition enters into a viscious cycle that impacts all the spheres of life and society. Malnutrition decreases resistance to infections thus more prevalence of diseases like pneumonia and diarrhoea in children. Deficiencies of micronutrients results into anaemic, stunted and weak children. Malnutrition results into poor mental development, resulting into poor learning and cognitive abilities. It results into poor school performance, increasing school dropouts and loss in IQ (5-11 points) of the children. It further results into low productivity, hence low income and poverty with estimated loss of 22 per cent in adult income. Ultimately malnutrition ends up with human capital loss and reduction in GDP to the tune of 11 per cent.

## Strategic Interventions taken by Ministry of Health and Child Welfare

Government of India has launched several programmes to achieve the global targets under SDG-2.2 goals. Various interventions have been taken up by the Government of India in child health care, which are briefly discussed under various subheadings.

### A. Newborn Health

☆ Essential new bore care and care around birth services are available at every delivery point.

☆ Facility based newborn care: Care of sick newborns is undertaken at every sub-district level in designated newborn stabilization units. At district level special newborn care units are available and at national level we have national neonatal intensive care units.

☆ Home based newborn and young child care: The two activities are merged together. Asha workers are important facilitators to ensure 5 visits of children at 3, 6, 9, 12, 15 months. Asha workers work for promotion of nutrition, mobilizing, counselling and seeking health care/services among the mothers at house hold level. Under this programme exclusive breastfeeding for six months is ensured. Adequate complementary feeding from six months and continued breast feeding up to two years of age is promoted. Growth monitoring and supplementation of iron and folic acid is regularly done. Full immunization for children is ensured, while use of ORS and zinc administration during diarrhoeal episodes is promoted. Age appropriate play and communication for children is taken care besides advocating proper hand wash practices.

✰ Comprehensive lactation Management Centres (CLMCs): Donor human milk centres/comprehensive lactation collection centres are being establishment at block level and funding for it is available. Boosting/promoting breastfeeding in lactating mothers, making available human mother milk wherever necessary is mandated with these centres.

## B. Nutrition Related Interventions

✰ Promotion of optimal infant and young child feeding practices under Mother's Absolute Affection (MAA) Programme

✰ Anemia Mukht Bharat programme aimed at micronutrient supplementation (Vitamin A, Iron, Folic Acid) and reducing anaemia burden. AMB is a 6 x 6 x 6 programme, with 6 targets, 6 mechanisms and 6 interventions for achieving AMB. 6 interventions are;

1. Prophylactic iron, folic acid supplementation
2. Periodic deworming of children, adolescents and pregnant women.
3. Intensified year round "behaviour change communication campaign" Solid Body Smart Mind.
4. Testing of anaemia using digital methods and point care treatment.
5. Mandatory provision of iron and folic acid fortified foods in public health programmes.
6. Addressing non nutritional causes of anaemia in endemic pockets, with special focus on malaria, haemoglobinopathies and fluorosis.

✰ Nutrition rehabilitation centres for management of children with severe acute malnutrition

## C. Other Important Interventions

✰ Management of Common Childhood illnesses like Diarrhoeal Diseases and Acute Respiratory Infections/pneumonia.

✰ Timely diagnosis of diseases, prompt referral, appropriate treatment and proper/justified use of antibiotics are important interventions.

✰ Elimination of diarrheal deaths through promotion of use of ORS and zinc.

✰ Capacity building programme.

### a. Integrated Management of Neonatal and Childhood Illness (IMNCI)

✰ The strategies for child health intervention focus on improving skills of the health care workers, strengthening the health care infrastructure and involvement of the community through behaviour change communication. The guidelines for health workers included;

✰ Improvements in the case-management skills of health staff through the guidelines on Integrated Management of Neonatal and Childhood illness.

☆ Improvements in the overall health system required for effective management of childhood illness.

☆ Improvements in family and community health care practices and involving them in health care process. In addition to these guidelines meant for peripheral health workers, Guidelines for training at other levels have also been developed.

**b.  Facility Based IMNCI (F – IMNCI) Training**

☆ It would provide the optimum skills needed by the Medical officers and Staff Nurses at the FRU's (First Referral Unit), thereby helps to address the acute shortage of Paediatricians at facilities.

☆ It focuses on providing appropriate inpatient management of the major causes of neonatal and childhood mortality such as asphyxia, sepsis, low birth weight, pneumonia, diarrhoea, malaria, meningitis and severe malnutrition in children at the FRU.

☆ The master trainers at state and district level are paediatricians from tertiary hospitals and medical colleges.

**D.  Rashtriya Bal Swasthya Karyakram (RBSK)**

Rashtriya Bal Swasthya Karyakram (RBSK) is an important initiative aiming at early identification and early intervention for children from birth to 18 years. It includes:

☆ Screening at delivery points for birth defects

☆ Screening at AWC and schools for 4 'D's *viz.*

    i. Defects at birth: Birth defects accounts for 9.6 per cent of all the newborn deaths that are being taken care through this RBSK programme.

    ii. Deficiencies: Various nutritional deficiencies affecting the preschool children

    iii. Diseases: Diarrhoea, Pneumonia, Anaemia

    iv. Development delays including disability. Developmental delays affecting at least 10 per cent of the children may lead to permanent disabilities including cognitive, hearing or vision impairment, and other diseases common in children *viz.* dental caries, rheumatic heart disease, reactive airways diseases *etc.*

☆ Establishment of DEIC: The District Early Intervention Centres are being established at District Hospital level across the country. The purpose of DEIC is to evaluate and manage all children below the age of 6 years by providing medical therapy, physiotherapy, psychological therapy, vision and hearing therapy, dental therapy, nutritional therapy, plaster therapy under one roof. It will also provide referral support for children requiring surgery or are above the age of 6 years provide referral support to children identified with health conditions.

# Chapter 10

# Importance of Agricultural Trade and Regional Cooperation for Food Security

**Joseph George**

*UNESCAP South and South-West Asia Office,
New Delhi*

Global food security is today more of an issue of access to food than its availability. Average dietary energy supply adequacy (ADESA), one of the key indicators used to measure sufficiency of food production/supply to meet the minimum nutritional requirement per capita, shows that globally food supply per capita is above sufficient levels.[1] And yet, more than 800 million people across the globe are undernourished, and malnutrition remains the leading cause of death in the world. This implies grossly uneven distribution of food resources and emphasizes the urgent need for correcting distortions in food markets. Recognizing the importance of trade in agriculture and food products for achieving the zero hunger target, by way of its potential impacts on correcting market distortions, target 6.a of Goal 2 (Zero Hunger) of the SDGs states as follows:

*Correct and prevent trade restrictions and distortions in world agricultural markets, including through the parallel elimination of all forms of agricultural export subsidies and all export measures with equivalent effect, in accordance with the mandate of the Doha Development Round.*

---

ADESA is calculated as the share of the Dietary Energy Supply (DES) (kcal/cap/day) in the Average Dietary Energy Requirement (ADER) expressed in percentages. Three year average (2017-19) index value for global aggregate stood at 119% (FAOSTAT).

Set as a means of implementation target, liberalized international trade in food products is envisaged as a key enabler for achieving the outcome targets of SDG 2. Trade enhances the prospects of achieving the zero-hunger target in many significant ways, both through the availability and accessibility channels. In terms of facilitating availability, open trade; (a) allows flow of food resources to flow from surplus to deficit markets, (b) enables dietary diversity by increasing the choices of food products available for consumption (important for nutritional fulfillment), (c) enhances marketability of agricultural outputs and thereby supports sustenance of food production, and (d) helps to reduce wastage and improve efficiency of food distribution. By way of improving accessibility, open trade; (a) helps to curb food price inflation and thereby support price stability and affordability, and (b) enables market expansion, increased export earnings, and supports livelihoods and purchasing power of farm sector dependents.

While creation of liberalized trade regimes through trade agreements dealing with both tariff and non-tariff barriers is an essential aspect of regional cooperation, its scope goes beyond the ambit of trade. International cooperation is needed to address a multitude of issues affecting agricultural production and consumption, especially those which are cross-border in nature such as climate change induced threats to agrarian eco-systems. In South Asia, with transboundary agroclimatic zones which are highly sensitive to climate change, regional cooperation is quintessential to find mitigative and adaptive solutions. Moreover, with similar challenges to food security policies, countries in the subregion can benefit from sharing R&D resources, technology and best practices.

## Challenges for South Asia on the way to Zero Hunger Target

South Asia has high stakes in the global fight against hunger under the framework of the Sustainable Development Goals (SDGs), with the unattained responsibility of uplifting the largest concentration of undernourished people in the world. About 257 million people were reported to have suffered from hunger in the subregion in 2019, accounting for well more than one-third (37 per cent) of the world's food deprived population. At 13.5 per cent in 2019, prevalence of undernourishment in South Asia is second only to Sub-Saharan Africa (22 per cent), and considerably higher than the world average (8.9 per cent). One in every three children affected by stunted growth due to poor nutritional intake belongs to South Asia. Adverse health conditions inflicted by malnourishment at early stages often last a lifetime, with serious implications for human resource development of the subregion. South Asia is also reported to have one of the highest incidences of what is referred to as hidden hunger or micronutrient deficiencies. As the numbers stack up, they leave no doubt as to the severity of food insecurity prevalent in the subregion.

On account of substantially higher population density compared to other subregions, South Asia has the highest hunger burden when the prevalence rate is translated into headcount of undernourished people, signifying that rate of reduction also has to be the highest for South Asia in order to achieve food security targets. It is therefore of alarming concern that the rate of reduction in number of food insecure people has been slowing down in the subregion. While the incidence of undernourishment diminished by a compound annual rate of close to 3 per cent between 2005-10 with an overall reduction by 13.7 per cent, the rate since then (2010-17) has been as low as 0.8 per cent with an overall reduction by only about 5 per cent.

Along with responsibilities of reaching the new higher targets of food security, the policy challenges in the way of meeting them are also on the rise. Hunger in South Asia is observed to be the result of a complex mix of factors acting through one or more of all the four pillars of food security, *viz.* availability, access, utilization and stability. Sensitivities of the agriculture sector in the subregional countries illustrate this. Slowdown in agricultural productivity in South Asia has a direct adverse impact on food availability, while also exerting its influence on food access by way of declining farm income. Livelihood dependency on agriculture has not declined proportionate to the dip in share of agriculture in the subregional economies. The slow and steady tilt in the terms-of-trade against agriculture, in favour of manufacturing and services, has had its effect on access of the agriculture dependents to food. Two broad categories of food security challenges facing South Asia, both of which necessitates agricultural trade and regional cooperation at a broader level are as follows:

## Development of Agrarian Systems (Production/Availability)

The surge in productivity following the green revolution peaked by 1990s (though there are inter-crop and inter-locale variations in productivity trends), though not before generating substantial improvements in sufficiency if food availability. However, some of the harmful effects of aggressive input intensification on agrarian system demands a renewed approach to productivity enhancement. Moreover, climate change threatens to thwart agrarian planning for greater productivity, while its triggers, onset, and possible impacts are not yet sufficiently understood. Addressing issues of productivity, requires cooperation in terms of R&D for ecological sustainability, cross-border investments in agricultural infrastructure and infusion of technological innovations in diverse areas of soil and water conservation, natural resource management, preservation of bio-diversity, irrigation, mechanization, fertilizers, seeds, pest management, post-harvest processing, storage and distribution *etc*.

## Enhancing Food Market Efficiency (Distribution/Access)

Market efficiency needs pertain simultaneously to both expansion of consumption basket and improvement of livelihoods. Often food grain security is equated with food security, ignoring micronutrient deficiency and its harmful health hazards. It is pointed out in the context of India that the share of consumer expenditure on cereals has reduced to account for a less than 40 per cent of total food consumption spending. Consumption basket has expanded considerably to include non-grain items, the demand for which is slated to rise at a pace in tandem with that of per capita income. As productivity shortfalls exist in the case of many non-grain food items, South Asia is inadequately prepared to meet the requirements of dietary diversity, without enhanced trade.

## Scope of Regional Trade in Agricultural and Food Products in South Asia

It will neither be feasible nor prudent to expect to fulfill South Asia's long-term food requirements with domestic production alone. With current trends and projections, meeting the future demand for food products, taking also into account the nutritional content and dietary fulfillment, the subregion would have to mainstream trade as a critical aspect of its strategies.

The current levels of intra-regional trade in food products in South Asia is found to be sub-optimal due to undue regulatory restrictions as well as procedural and infrastructural barriers to trade. Trade in agriculture is observed to be afforded more protection compared to non-agriculture sectors. Intra-regional food trade is also highly concentrated, as over 90 per cent regional food exports is constituted by a few product categories such as cereals, dairy products, sugar, vegetable oils, tea and spices. Country participation is also skewed as India and Pakistan account for almost 77 per cent of overall intra-regional food exports in the subregion. The share of processed food in intraregional trade has been observed to be considerably lower compared to non-processed food. This may be a validation of fact that quality standards related issues dampen the subregions trade as imposition of quality standards is observed to have more severe effect on the processed food sector than the non-processed food sector.

Import of nearly all agricultural outputs, including livestock and processed food products are subjected to some kind of sanitary and phytosanitary (SPS) certification and import permit. Often the procedural inefficiencies with respect to testing, certification and inspection are found to be greater trade deterrents than the standards themselves. Trade in food commodities, especially non-processed food, is also dependent on timely delivery due to short shelf-lives, and therefore relies heavily on facilitative trade infrastructure. Longer turnaround times due to transshipment and congestion at border points, and absence of testing and quarantine facilities and cold storage chains, at the South Asian land customs stations affect trade to a large extent.

With the exception of the land-locked countries of Afghanistan, Bhutan and Nepal, whose top exporters are South Asian partners, trade in agricultural products of the subregional countries are predominantly oriented towards external trading partners. For most of the South Asian countries their major import sources and export destinations of food and agricultural products are outside the region, with some accounts putting such external trade as constituting more than 90 per cent of the subregion's overall agriculture trade. There are substantial gains to be reaped by merely internalizing costly external trade to the extent possible.

A study commissioned by UNESCAP explores possibilities of replacing external imports of important food items with imports from South Asian trading partners where supply capacity for the same exists within the subregion. The study used the concept of Additional Market Access Frontier (AMAF) – matching high supply potential and high import demand within the region - to assess the potential for imports of important food items by South Asian countries from subregional trading partners, possibly at a cheaper price. This exercise identifies a list of 397 food products with a combined internal trade potential of about US$ 10 billion per annum, amounting to about 40 per cent of the current aggregate intra-regional trade of the subregion. Realizing this potential can result in huge benefits in terms of price and supply stability.

## Priorities for Liberalizing Agricultural Trade

South Asian countries should cooperate on a host of regulatory and trade facilitation reforms for achievement the full potential of intra-regional food trade. Agriculture is afforded more trade protected than non-agriculture in the subregion. Even though the level of tariff protection of food products trade has come down through the SAFTA TLP (South Asia Free Trade Area Trade Liberalization Programme) and various bilateral trade arrangements, a large number of products are still contained in their sensitive lists. The study by UNESCAP provides a potential list of commodities to prune the sensitive lists ad apply preferential tariff rates. However certain procedural requirements are also required, besides tariff liberalization, including simplified rules of origin (RoOs) and quality compliance, quarantine, testing and certification.

On the infrastructure side, the priority is to ensure that the land customs stations are equipped to the adequacies of handling food products. Procedural reforms should focus on minimizing the cost of compliance with standards through Mutual Recognition Agreements (MRAs). A host of existing policy instruments available under SAFTA and institutions set up under the SAARC framework can be used to expedite reforms in this regard. Initiatives that are underway at SARSO (South Asian Regional Standards Organization) for efficient and transparent regional standards can exert a substantial impact on food trade in the subregion

## Taking Regional Cooperation for Food Security Beyond Agricultural Trade

The scope of regional cooperation should reach beyond the ambit of trade. Slowdown in South Asia's yearly rate of improvement in the food security situation in recent times, at a time when nothing short of an unprecedented acceleration in progress would suffice, is an indication that the subregion now confronts what may be called its bottom territory of food deprivation or that critical domain of extreme hunger which is the most difficult to tackle. It implies that conventionally followed policy responses are likely to be found wanting. In order to be effective, future course of actions should be founded on innovative solutions capable of addressing the new challenges, and the chances of finding such solutions are much higher when they are collectively sought.

Regional cooperation for food security is hardly a new concept in South Asia. In fact, sustainable agriculture and food security were among the most important priorities in the early years of SAARC. The Agreement Establishing South Asia Food Security Reserves was signed as early as 1987. The 15th SAARC Summit adopted what is referred to as the Colombo Statement on Food Security, identifying six thematic areas, including food production, investments in agro-industries, agricultural technologies, market interventions, and management of risk. Successive rounds of SAARC Summits have provided mandate for cooperation in wide range of topics related to food security, including, among others, the call for establishing regional food and seed reserves. The directives promised cooperation in areas ranging from agricultural productivity to technology, trade, food safety, price control, bio-diversity, climate change and natural resource management. However, regionally coordinated policy responses in each of these areas are yet to take definitive shapes.

Comprehensive regionally coordinated action plans are required to save such sensitive ecosystems from floods and soil erosion, storm surges and extreme cyclonic events. One if the main areas in which regional cooperation can deliver over unilateral measures to a large extent is by way of integrating the use of climate forecasts into cropping decisions. Regional approaches would help to harness greater resource mobilization potential, multi-stakeholder partnerships and alliances to tackle complex issues of transboundary environmental risks which are long-term in nature, the solutions for which are hardly possible exclusively through unilateral measures at the national level. Substantial increase in budgetary commitments are required to fulfil the new and reconfigured public policy targets of food security. Besides the obvious advantages of resource mobilization, the single most important benefit of regional cooperation would be that of greater allocative efficiency. Some of the key areas of joint interventions for South Asia, valued against the subregions challenges, are with respect to building resilience to climate change, trade liberalization, operationalization of the SAARC food and seed banks, technical capacity building and knowledge sharing, improving governance through regional institutions, management of natural resources *etc*.

A 10-point regional action agenda on food security emerging from UNESCAP's work on this topic covers:

1. Cooperation for combatting climate change,
2. Regional trade liberalization in agriculture,
3. Operationalization of regional food reserves (SAARC Food Bank),
4. Leveraging technology,
5. Sharing of good practices,
6. Building regional institutions,
7. Coordinated positions in multilateral forums on agriculture and food security,
8. Addressing transboundary outbreak of livestock diseases,
9. Strengthening food safety standards, and
10. Collective measures for management of shared natural resources.

# Chapter 11

# Sustainable Agricultural Research and Technology: Needs to Address Food Security

Zahoor A. Ganie

*Scientist Herbicide Discovery, FMC Corporation, New York, DE, USA*

According to the Food and Agriculture Organization (FAO) there are approximately 820 plus million people suffering from hunger around the world. The paradox is that though there is enough food to feed the current world population still millions are faced with hunger because one-third of current food production is lost after harvest at the farm level or wasted in the food distribution channels. Some of the major reasons for such huge losses or wastage of food is lack of proper storage, inadequate connectivity/transportation system, marketing strategies, consumer attitude and dearth of awareness on the impact of food wastage. Reducing the post-harvest losses and wastage of food should be the priority to address the issue of hunger at present. Post-harvest processing of fruits, vegetables and grains holds great promise to reduce the food loss. There is need for more investment on applied post-harvest management research/technologies, policy change to focus on infrastructure development including cold storage, refrigerated transportation, and efficient distribution systems *etc*.

Global population is projected to reach 9 plus billion by 2050 and that will require food production to increase by 70 per cent over the current level. Achieving such a massive increase in food production will be challenging due to mounting pressure on diminishing natural resources including land, water, and energy. Besides, climate change and global warming will further make the food production complicated if not impossible in future. Therefore, sustainability has become more important than ever before to attain global goals of food production needed to feed the future populations. Sustainable agriculture emphasizes on meeting the needs of current population by using the resources in an efficient manner without compromising the ability of future generations to meet their needs of food and quality environment. Sustainable agricultural practices include, but are not limited to, conservation agriculture, high yielding resource efficient varieties with fortified food quality, improving the water use efficiency through technologies like drip irrigation, alternate wetting and drying in rice paddies, and reducing the environmental foot-print of agriculture by cutting down the green-house gas production *etc*.

Conservation agriculture is based on key principles including minimum tillage, retaining a permanent soil cover and crop diversification. The concept of minimum tillage involves placement of seed and fertilization in precisely opened rows in a single trip while leaving the soil undisturbed in-between the rows. Minimum tillage is crucial to reduce the soil erosion, preserving the soil structure, avoiding soil compaction, and reducing the fuel use, cost, and pollution. Similarly, retaining a ≥30 per cent permanent soil cover improves soil organic matter, enhances soil-water infiltration, and helps in weed suppression. Likewise, crop diversification is beneficial for homogenously tapping soil resources (moisture, nutrients), fostering diverse soil life (flora and fauna) and breaking the insect/disease and weed cycles.

New crop cultivars (varieties and hybrids) with improved input use efficiency such as water/nutrient use efficiency, better light interception, drought tolerance, tolerance/resistance to insects and diseases and competitive ability against weeds, plasticity towards unpredictable weather elements and better nutritional quality per unit of produce are needed. Cropping systems should be revised to include high resource efficient crops and sustainable agronomic practices for better productivity, higher profitability and reducing the environmental footprint of agriculture. For example, research and technology to develop efficient systems of subsurface and drip irrigation systems in place of current less efficient surface irrigations. Similarly, direct-seeded rice is a promising technology to reduce the demand on water, labor, cut down methane emission while increasing the profit margins of rice-based cropping systems. However, there are certain challenges to direct seeded rice at present including high weed pressure and lack of standardization of varieties and other agronomic practices.

The greatest challenges for agriculture are variability in the soil in terms of fertility, moisture, soil health and instabilities in the weather conditions including fluctuations rather unpredictability in precipitation, temperature, humidity, and wind. Huge progress is underway in technologies such as remote sensing, weather monitoring, and drones endowed with hyperspectral imaging cameras *etc.* to collect data needed to understand the variability in soil and weather conditions. This information may be used to develop timely predictions to support decision making processes of growers related to input allocation, pest management and other production practices. Precision agriculture is an emerging example of this technology that utilizes data on spatial and temporal variations in the agroecosystem to make accurate agronomic decisions without wasting resources and time. In future, the information on different aspects of agroecosystem will be transformed into more valuable forms that can be used in different innovative known or yet to be known ways. It is expected that data on different aspects of agroecosystems will make it possible to supply inputs on yield target basis to each single farm/grower and to deliver customized solutions for pest management *etc*. An example of promoting sustainable agricultural practices is the recent program launched in north America by a leading industry to incentivize agricultural practices that increase carbon sequestration by offering economic rewards referred as carbon credits. However, there is still more research needed to understand the potential use and standardization of this data.

In summary, enormous increase in food production to feed the future population must come from efficient use of dwindling resources (land, water) while reducing the environmental footprint of the agriculture and avoiding the adverse impact of climate change. Innovative research and technologies are needed to optimize the agricultural production processes for improving the productivity, profitability, socio-economic feasibility, and overall sustainability of the agricultural systems.

# REFERENCES


1.  How to feed the world 2050. http: //www.fao.org/fileadmin/templates/wsfs/docs/expert_paper/How_to_Feed_the_World_in_2050.pdf

2.  https: //www.wri.org/blog/2018/12/how-sustainably-feed-10-billion-people-2050-21-charts

3.  Suraj B and Behera UK (2014) Conservation agriculture in India – Problems, prospects, and policy issues. International Soil and Water Conservation Research, 2: 1-12

4.  www.fao.org/ag/ca/

5.  Singh Y, Singh VP, Chauhan B, Orr A, Mortimer AM, Johnson DE, Hardy B, editors. 2008. Direct seeding of rice and weed management in the irrigated rice-wheat cropping system of the Indo-Gangetic Plains. Los Baños (Philippines): International Rice Research Institute, and Pantnagar (India): Directorate of Experiment Station, G.B. Pant University of Agriculture and Technology. 272 p.

6.  https: //ipmworld.umn.edu/hutchins

( Chapter 12 )

# Crop Biofortification for Achieving Zero Hunger by 2030


Z.A. Dar, B. Kumar, R. Munshi, S. Naseer, F. Rasool,
A.A. Lone, N.S. Khuroo, S.A. Dar, I. Abidi,
G. Ali, A.M. Iqbal and A. Khan

*DARS, Rangreth-SKUAST-Kashmir*


## Introduction

Agriculture is the mainstay of most of people globally, and especially crucial in rural areas where the majority poor and vulnerable population live [1]. Agriculture, not only provides food and nutrition to support human life, but also constitutes a major source of income to many people around the world [1–3]. Moreover, agriculture determines food availability and access, affects food price as well as community health and nutrition status. According to Food and Agriculture Organization [4] about 815 million people are chronically hungry and go to bed without food. Recently, hunger has increasing from 777 to 815 million people in 2015 and 2017, respectively [4]. At the same time, micronutrient deficiencies have become a global burden, as every third person is malnourished, reflecting agriculture, food and nutrition systems out of balance [5,6].

The global hidden hunger statistics in 2016, indicated 11 per cent of the population being undernourished, 22.9 per cent children stunted and 33 per cent women at reproductive were anemic [5]. The situation is worsening in sub-Sahara Africa, South-eastern and Western Asia [4]. Whereas, sub-Sahara Africa account for 90 per cent of the burden with 40 per cent stunting, 50 per cent vitamin A deficiency and 30 per cent anemia among pre-school children [7,8]. Micronutrient deficiencies are mainly due to limited access to food rich in vitamins and minerals in the diet, important to support human health [9]. Hidden hunger is the collective term that describes a condition of undernutrition where the body lack vitamins and minerals that keep people healthy [8]. Micronutrients of public health concern includes; iron, zinc, iodine, vitamin A and B9 among others [10]. Consequences of micronutrients deficiencies to human health ranges from low birth weight, poor cognitive potential, weak immune system, anemia, stunting, to incidences of adulthood nutritional non-communicable diseases such as *diabetes mellitus* [11,12]. Collectively, hidden hunger causes irreversible damage to individuals. Yet, one in four children under the age of five years are at increased risks of impaired cognitive and physical performance, stunted and prone to infections [4]. Iron deficiency leads to anemia and stunted growth which contributes to reduced mental development, poor cognitive potential and reduced physical performance in children [9]. Zinc deficiency is indicated with stunting, poor immune system, diarrhea and increased risks of respiratory diseases affecting 2 billion people worldwide [13,14]. Vitamin A deficiency results in poor health, reduced cognitive potential, weak immune, dry eyes and night blindness, a disorder like xerophthalmia [15,16]. Iodine deficiency disorder health effects include; goiter, cretinism, growth retardation, hypothyroidism, and increased pregnancy loss and infant mortality [17].

Moreover, insufficiency maternal iodine intake may result in neurological and cognitive disorders in children. Folate deficiency during pregnancy is linked to neural tube defect in fetus, diminished deoxyribonucleic acid (DNA) methylation, neuropsychiatric manifestation and cognitive problems [18,19]. Undernutrition, overweight, obesity and nutritional non-communicable diseases throughout the life course, coexisting in different countries, communities and household presents malnutrition double burden challenges [20,21]. For instant, Dietz [21], writes: 'increased use of processed complementary food for children could provide calories without micronutrients, and thereby increase the likelihood of obesity, undernutrition, and stunting and obesity in the same children.' To address the challenges, United Nations member states are committed to end hunger and any form of malnutrition by 2030 under global Sustainable Development Goals (SDGs), SDG2 '**zero hunger**'[22]. It provides a platform for the United Nations system, governments, private sectors, and other stakeholders for collectively impact agriculture, food and nutrition security in a sustainable manner.

In response, several nutrition sensitive as well as specific interventions have been introduced to alleviate hidden hunger which includes; micronutrient supplementation, food fortification, crops biofortification, promoting dietary diversification, better public health system and diseases control measures [23]. However, food fortification programs proved to be expensive not reaching the majority and people with little access to processed food, especially those scattered and isolated in the rural areas in developing countries [24]. Also, the nutrient supplements are in a medical formulation delivered in form of syrup or tablets following diagnosis or being certain that, an individual has a deficiency [25].

Crop biofortification provides a scientific approach through agronomic, selection breeding and genetic engineering of plant to enhance micronutrients content [26]. It is a food-based approach that offer wide range of nutrients that are essential for human health. Some of biofortified crops include: iron and zinc biofortified beans and millet; β-carotene enhanced orange fleshed sweet potato (OFSP), yellow maize and cassava, orange banana and golden rice among others [27]. Quite a number of agriculture innovations for biofortified crops had been conducted in different region, but little is known about nutrient bioavailability and community perception related to food habits and custom. As the results, people are suffering from hunger and hidden hunger while agriculture has been the best feasible and sustainable means to providing food and nutrition. Therefore, this review intends to create awareness on the potential of biofortified crops and inform the community, researchers and policy makers on unintended narrowed food diversity following several interventions which promote biofortified crops in order to engender better agriculture policies that promote nutrition.

## Crop Biofortification

Biofortification is the process by which the nutritional quality of food crops is improved through agronomic practices, conventional plant breeding and/or modern biotechnology techniques [28]. Biofortification differs from food fortification in that crop biofortification aims to increase nutrients level in the crop during plant growth than manual means during food processing [28]. Basically, biofortification intends to enhance micronutrients including trace elements of staple crops and provide a public health benefit with minimal health risks. It is a sustainable intervention that use dietary approach in an initiative to combat hidden hunger [29]. It takes advantages of staple food to reach the majority of undernourished population in the remote areas who depends on agriculture for their food and nutrition. Biofortification may therefore provide a way to reach populations where food fortification and micronutrients supplementation activities may be difficult to implement and/or limited [28].

Targeted crops include; rice, sweet potato, potato, maize, wheat, common beans, sorghum, tomato, cassava, banana, barley, soy beans, lettuce and carrot. Agronomic biofortification is achieved by application of nutrient rich fertilizers to foliage or soil to enhance micronutrients concentration in crop edible part hence increased micronutrients intake by consumers [30,31]. Largely, agronomic biofortification depends on the bioavailability of nutrients from the soil to plant. The use of selenium and zinc rich fertilizers had been successful in increasing selenium and zinc in rice and wheat, respectively [29,32]. Biofortification through conventional plant breeding has successfully increased the level of nutrients in the progeny plant [28]. Moreover, increased levels of iron, zinc, vitamin A and selenium has been achieved by conventional breeding method in most of plants [33,34].

Although, crop biofortification has been achieved through agronomic practices and conventional breeding, in some crops where target micronutrient does not occur naturally, transgenic plant breeding is opted [35,36]. Genetic modification and molecular techniques are used in transferring a specific trait; ability to synthesis micronutrients from a donor organism to recipient organism such as rice, so that it expresses the trait; enhances micronutrients in the edible portion through increased efficiency of biochemical pathways [28,37]. For instance, transgenic iron, zinc and β-carotene biofortified rice have been genetically engineered, representing one of the successful stories of transgenic plant breeding [38]. Other biofortified crops include: pro-vitamin A enhanced soy bean; transgenic enhanced β-carotene and lycopene tomato; enhanced resveratrol antioxidant capacity in apple; enhanced methionine-amino acid in common beans; transgenic β-carotene, ascorbate and folate enhanced maize; transgenic β-carotene potato; enhanced calcium bioavailability in carrot; transgenic lettuce with improved iron content; and pro-vitamin A enhanced yellow cassava, OFSP, golden rice and banana [27,37]. Additionally, transgenic breeding can also be used for incorporation of genes responsible for reduction of anti-nutritional factors, thereby enhancing micronutrients bioavailability [9,39]. While these transgenic biofortified crops have proved nutritional potential, they have not been introduced in large part due to high risks and regulations imposed on genetically modified crops [40,41].

## The Biofortification Process

Biofortication requires experts from different fields to work together. Plant breeders explore the full spectrum of crop genetic diversity, especially seed banks, to first identify nutrient-rich germplasm, or lines, of food crops that can be used to breed more nutritious varieties. These lines are then crossed with established high-yielding lines to breed new crop varieties that not only have higher amounts of a desired nutrient, but also are high yielding and competitive with other non-biofortified varieties. Plant breeders can use both conventional plant breeding and transgenic methods to reach their breeding targets. Nutritionists must determine the additional amount of a nutrient a food crop must provide to measurably improve nutrition when that crop is harvested, processed or cooked, and eaten. To do so, nutritionists must account for:

1. Nutrient losses after the crop is harvested (nutrients can degrade substantially during storage, processing, or cooking).
2. The amount of the nutrient that the body actually absorbs from the food (bioavailability).
3. The amount of the staple food actually consumed on a daily basis by age and gender.

These data are then used to set breeding targets for specific nutrients. Once these new crop lines have been bred, they are field tested by a national agricultural system in multiple locations in target regions where the crop will be grown. This ensures the crops perform well and maintain their nutritional profile, which can be affected by the growing environment. The most promising lines are selected for further testing and eventual release as new varieties through public channels, the private sector, or both.

For farmers to grow the biofortified crops, the nutrient-rich varieties should be economically remunerative. The speed of adoption of the seed will depend upon the efficiency of the seed distribution system, and the support of public extension services. For consumers to consume, besides taste, texture, and visual appearance, awareness about malnutrition, easy availability, and access of cereals at an affordable price are crucial. It may be essential to understand that consumers' acceptance influences farmers' adoption decisions and vice-versa. Low acceptance by the consumers will lower market price for farmers, and low adoption by farmers will make the product unaffordable for the poor consumers.

Special consideration should be given to crops whose color or taste is changed by improved nutrient content. To date, this has been the case when crops such as sweet potato, cassava, and maize have been enhanced with vitamin A. These crops turn from a typical white or pale yellow to a deeper yellow or orange in color due to the higher levels of beta-carotene (a precursor to vitamin A) they now contain. This orange color can be an asset in helping consumers identifies more nutritious varieties. Special consideration should be given to deliver biofortfied staple food crops with one or more of the most limiting nutrients in the diets of the poor like vitamin A, zinc, and iron.

## Evidence Linking Crop Biofortification, Food and Nutrition

### Vitamins

A number of interventions have been undertaken to achieve crop vitamins enhancement. Provitamin A ($\beta$-carotene) crop biofortification is among popular successful biofortification programs. $\beta$-carotene is converted in the body to retinol, a form of vitamin A used in the body. It has been demonstrated that increasing provitamin A intake through consuming $\beta$- carotene enhanced crops, results in increased vitamin A body stores in different human age groups [42,43]. For instance, in Uganda and Mozambique intake of OFSP improved vitamin A status in children and women and decreased the prevalence of low serum retinol [44,45]. In Zambia, 5-7 years old children fed with $\beta$- carotene enhance orange maize showed an increased total body store of vitamin A compared with the control group [46]. Children 5-13 years old and women of child bearing age consumed $\beta$-carotene biofortified yellow cassava in Kenya and Nigeria, showed a significant improvement in vitamin A status indicated as serum retinol [47– 49]. Also, consumption of $\beta$-carotene biofortified crops resulted in improved visual performance, increased serum $\beta$-carotene and liver retinol concentration in marginal vitamin A deficient children in a study conducted in Zambia [42,46].

Vitamin B6 biofortified transgenic cassava showed 4-14 folds and 3-15 folds increase in vitamin B6 concentration in leaves and storage starchy roots, respectively [50,51]. Similarly, vitamin B1 (thiamine) and B9 (folate) has been enhanced in rice, although affected by processing such as polished rice of which normal practices people consume polished rice [52]. Other crops enhanced with vitamin B through transgenic methods include; tomato, lettuce, maize and potatoes. Moreover, Mène-Saffrané and Pellaud [53], reported development achieved to enhance vitamin E in soybean and maize.

### Minerals

Crop iron, zinc, iodine and selenium biofortification has been achieved in rice, wheat, maize, finger millet and common beans to mention few. Iron biofortified beans in Rwanda showed a significant increase in hemoglobin, total blood iron and improved cognitive potential when iron-depleted women (18-27 years old) were fed with iron biofortified beans [54,55]. Iron biofortified finger millet in India, showed a significant increase in serum ferritin and improved physical performance in school children, adolescent boys and girls who were iron deficient [56]. Moreover, consumption of biofortified crops reported to reduce prevalence and duration of diarrhea, incidences of infection and reduced the likelihood of marginal micronutrient deficiencies in children under five years in Mozambique [57,58]

## Advantages of Biofortification

Dietary diversity is the ultimate long-term solution to minimizing hidden hunger. This will require substantial increases in income for the poor so they are able to afford more nutritious nonstaple foods such as vegetables, fruits, and animal products. Biofortification can be effective in reducing hidden hunger as part of a strategy that includes dietary diversification and other interventions such as supplementation and commercial fortification. Biofortification has four main advantages when applied in the context of the poor in developing countries. First, it targets the poor who eat large amounts of food staples daily. Second, biofortification targets rural areas where it is estimated that 75 per cent of the poor live mostly as subsistence or smallholder farmers, or landless laborers. These populations rely largely on cheaper and more widely available staple foods such as rice or maize for sustenance. Despite urbanization and income growth associated with globalization, diets of the rural poor will continue to be heavily based on staple foods like cereals and tuber crops in many regions. Expected increases in food prices, exacerbated by climate change, are likely to increase this reliance on staple foods. Third, biofortification is cost effective. After an initial investment in developing biofortified crops, those crops can be adapted to various regions at a low additional cost and are available in the food system, year after year. Fourth, because this strategy relies on foods people already eat habitually, it is sustainable. Seeds, roots, and tubers can usually be saved by farmers and shared with others in their communities. Once the high-nutrition trait is bred into the crops, it is fixed, and the biofortified crops can be grown to deliver better nutrition year after year—without recurring costs.

## Biofortification: Limitations and Challenges

Biofortification requires high adoption by both farmers and community. Indeed, the visibility of traits and infrastructures are critical to technology adoption [59,60]. Biofortified crops with visible traits such as orange-fleshed sweet potatoes and golden rice requires that producers and consumer accept these changes in addition to claimed nutritional potentials [59]. Enhanced β- carotene which intensify the color to yellow and dark orange become problematic for acceptance for many in Africa, where white-fleshed sweet potatoes, cassava, maize and rice are preferred [59,61]. Also, enhanced β-carotene in cassava roots resulted in changed dry matter content in a study conducted in Nigeria [62]. For instance, yellow maize was negatively perceived as food aid which was suitable for animal feed during time of hunger in Zambia and most of the African countries [63]. Also, the contribution of yellow maize to egg yolks, animal fat and poultry skin yellow coloration has attracted their use as animal feed [64].

Promising as it is, biofortification faces limitations and challenges. First, biofortification requires a paradigm shift. Agricultural scientists need to add nutrition objectives to their breeding programs, in addition to standard goals such as productivity and disease resistance. Plant breeders must then work closely with nutritionists to develop breeding targets for nutrients. Nutritionists and health professionals also need to accommodate agriculture-based approaches in their toolbox along with clinical interventions. Agricultural science and nutrition are compartmentalized disciplines that must integrate for biofortification to succeed. Second, biofortification will be widely adopted only when proponents show these new foods improve nutrition. Most biofortified crops are still in the development pipeline. However, one biofortified staple food crop that has been successfully released is the orange (or orange-fleshed) sweet potato. As other crops follow, nutritionists will be able to build a body of evidence that biofortification is a viable agriculture based intervention to improve nutrition. Third, the amounts of nutrients that can be bred into these crops are generally much lower than can be provided through fortification and supplementation. However, by providing 30–50 per cent of the daily nutrient requirement, biofortified crops can significantly improve public health in countries where hidden hunger is widespread (poor consumers in most cases will already be consuming 50 per cent of requirements). Transgenic approaches can be used to improve the nutrient content of crops where natural variation in germplasm is limited. However, transgenic crops also face more regulatory hurdles compared to their conventionally bred counterparts. Whether conventionally or transgenically bred, biofortified crops should shift significant numbers of people that are receiving a little less than their estimated nutrient requirement, into a state of nutritional adequacy, for that nutrient. Fourth, nutritionists now focus on the 9-24 month age group, when micronutrients are crucial for healthy development. Infants consume relatively low amounts of staple foods and yet have relatively higher micronutrient requirements, making biofortification's contribution to micronutrient adequacy in this group limited. There are exceptions; due to the particularly high vitamin A content of many OSP varieties, regular consumption of these by the mother could contribute substantially to vitamin A intakes of breastfed children 6–23 months of age. In Mozambique and Uganda, a HarvestPlus project also showed substantially improved vitamin A intakes from OSP in children aged 6–35 months. However, researchers need to better understand biofortfication's potential impact on the -9-to-24-month age group through the mother's micronutrient status going into pregnancy, when her micronutrient requirements substantially increase. This micronutrient status could be better for mothers who have consumed biofortified crops from adolescence, or even earlier.

## Conclusion

Changes in agriculture technologies, food systems and dietary patterns has led to a narrowed food choice to provide nutrients. The increased consumption of highly starch biofortified food crops in many countries, signal a shift from traditional diverse food diet. However, biofortified crops are either with visible or invisible traits and sometimes changed taste and dry matter content challenging consumer acceptability. Thus, a trend that possibly explains the continuing coexistence of multiple form of malnutritional; undernutrition, hidden hunger, overweight, obesity and nutritional noncommunicable diseases within communities and even the same households. But, if crop biofortification programs include nutritional education that promote food diversity, could make it a reality. For instance, the bioavailability of β-carotene depends on oils/fat; iron on vitamin C; and other nutrient-nutrient interactions in the diversified diet. Therefore, it is high time to integrate nutrition education, consumer food and dietary habits, custom and culture to create a need of biofortified crops such that when the project ends, people still see the need to cultivate traditional and biofortified crops, at the same time embraces dietary diversification.

# REFERENCES


1.  Bruinsma J. World Agriculture: Towards 2015/2030: an FAO study. London: Routledge. 2017;444.

2.  James A and Zikankuba V. 2017. Postharvest management of fruits and vegetable: A potential for reducing poverty, hidden hunger and malnutrition in sub-Sahara Africa. Yildiz F, editor. Cogent Food Agric [Internet]. 3(1)

3.  Gödecke T, Stein AJ, Qaim M. 2018. The global burden of chronic and hidden hunger: Trends and determinants. Glob Food Sec [Internet].17: 21–9.

4.  Food and Agriculture Organization. SOFI 2017 - The State of Food Security and Nutrition in the World [Internet]. How close are we to # ZeroHunger? 2017 [cited 2018 May 29].

5.  FAO, IFAD, UNICEF, WFP, WHO. Building Resilience for Peace and Food Security. In: The State of Food Security and Nutrition in the World [Internet]. Rome; 2017 [cited 2018 May 29].

6.  FAO. 2018.Transforming Food And Agriculture To Achieve The SDGs: 20 interconnected actions to guide decision-makers [Internet]. Rome.

7.  von Grebmer K, Saltzman A, Birol E, Wiesman D, Prasai N, Yin S, *et al.*, 2014. Global hunger index: The challenge of hidden hunger. In: IFPRI books [Internet]. International Food Policy Research Institute (IFPRI).

8.  Muthayya S, Rah JH, Sugimoto JD, Roos FF, Kraemer K, Black RE. 2013. The global hidden hunger indices and maps: An advocacy tool for action. Noor AM, editor. PLoS One [Internet]. Jun 12 [cited 2018 May 29]. 8(6): e67860.

9.  Vasconcelos MW, Gruissem W, Bhullar NK. 2017. Iron biofortification in the 21st century: setting realistic targets, overcoming obstacles, and new strategies for healthy nutrition. Curr Opin Biotechnol. 44: 8–15.

10. Burchi F, Fanzo J, Frison E. 2011. The Role of Food and Nutrition System Approaches in Tackling Hidden Hunger. Int J Environ Res Public Health. 8(2): 358–73. Available: http://www.mdpi.com/1660- 4601/8/2/358

11. Tulchinsky TH. 2010. Micronutrient deficiency conditions: Global health issues. Public Health Rev. 32(1): 243–55.

12. Bhandari S, Banjara MR. 2015. Micronutrients Deficiency, a Hidden Hunger in Nepal: Prevalence, Causes, Consequences, and Solutions. Int Sch Res Not [Internet].2015: 1–9.

13. Prasad AS. 2013. Discovery of human zinc deficiency: Its impact on human health a disease. Adv Nutr. 4(2): 176–90.

14. Gibson R.S. 2012. Zinc deficiency and human health: etiology, health consequences, and future solutions. Plant Soil [Internet]. 361(1–2): 291–9.

15. Wirth J, Petry N, Tanumihardjo S, Rogers L, McLean E, Greig A, *et al.*, 2017. Vitamin A Supplementation Programs and Country- Level Evidence of Vitamin A Deficiency. Nutrients [Internet].9(3): 190.

16. Martini S, Rizzello A, Corsini I, Romanin B, Fiorentino M, Grandi S, *et al.*, 2018. Vitamin A deficiency due to selective eating as a cause of blindness in a high-income setting. Pediatrics. 141(Suppl 5): S439–44.

17. Pearce EN. 2017. Iodine Deficiency Disorders and Their Elimination [Internet]. Cham: Springer International Publishing; [cited 2018 May 29]. Available: http://link.springer.com/10.1007/9 78-3-319-49505-7

18. World Health Organization. 2015. Optimal Serum and Red Blood Cell Folate Concentrations in Women of Reproductive age for Prevention of Neural Tube Defects [Internet]. Geneva: World Health Organization; [cited 2018 May 30].

19. Fenech M. 2012. Folate (vitamin B9) and vitamin B12 and their function in the maintenance of nuclear and mitochondrial genome integrity. Mutat Res Mol Mech Mutagen [Internet]. May 1 [cited 2018 May 30]. 733(1–2): 21–33. Available: https://www.sciencedirect.com/science/article/pii/S0027510711002934

20. Prentice AM. 2018. The Double Burden of Malnutrition in Countries Passing through the Economic Transition. Ann Nutr Metab [Internet]. 72Suppl 3(3): 47–54.

21. Dietz WH. 2017. Double-duty solutions for the double burden of malnutrition. Lancet (London, England). 390(10113): 2607–8.

22. Amoroso L. 2018. Post-2015 Agenda and Sustainable Development Goals: Where Are We Now? Global Opportunities to Address Malnutrition in all Its Forms, Including Hidden Hunger. In: Hidden Hunger: Strategy to Improve Nutrition Quality [Internet]. Karger Publishers. 45–56.

23. Ruel MT, Alderman H. 2013. Nutrition-sensitive interventions and programmes: how can they help to accelerate progress in improving maternal and child nutrition? Lancet [Internet]. 382(9891): 536–51.

24. Thompson B, Amoroso L, 2010. Combating Micronutrient Deficiencies: Food-based Approaches [Internet]. CABI;

25. Zimmermann MB, Hurrell RF. 2007. Nutritional iron deficiency. Lancet [Internet]. 370(9586): 51120.

26. Briat J-F, Dubos C, Gaymard F. 2015. Iron nutrition, biomass production, and plant product quality. Trends Plant Sci [Internet]. 20(1): 33–40.

27. Garg M, Sharma N, Sharma S, Kapoor P, Kumar A, Chunduri V, *et al.*, 2018. Biofortified Crops Generated by Breeding, Agronomy, and Transgenic Approaches Are Improving Lives of Millions of People around the World. Front Nutr [Internet]. [cited 2018 May 11]. 5: 12.

28. WHO. 2016. WHO | Biofortification of staple crops. e-Library Evid Nutr Actions [Internet]. [cited 2018 Jun 8]; Available: http: //www.who.int/elena/titles/bio fortification/en/

29. de Valença AW, Bake A, Brouwer ID, Giller KE. 2017. Agronomic biofortification of crops to fight hidden hunger in sub-Saharan Africa. Glob Food Sec [Internet].12: 8–14. Available: https: //www.sciencedirect.com/science/article/ pii/S2211912416300487

30. Klikocka H, Marx M. 2018. Sulphur and Nitrogen Fertilization as a Potential Means of Agronomic Biofortification to Improve the Content and Uptake of Microelements in Spring Wheat Grain DM. J Chem [Internet]. [cited 2019 Feb 10]. Available: https: //www.hindawi.com/journals/jchem/2018/9326820/ abs/

31. Saeid A, Jastrzêbska M. 2017. Agronomic biofortification as a key to plant/ cereal fortification in micronutrients. In: Food Biofortification Technologies [Internet]. CRC Press.1–60. Available: https: //www.taylorfrancis.com/ books/9781498756600/chapters/10.1201/978135122850-1

32. Lidon FC, Oliveira K, Ribeiro MM, Pelica J, Pataco I, Ramalho JC, *et al.*, 2018. Selenium biofortification of rice grains and implications on macronutrients quality. J Cereal Sci [Internet].81: 22–9.

33. Khush GS, Lee S, Cho J-I, Jeon J-S. 2012. Biofortification of crops for reducing malnutrition. Plant Biotechnol Rep.6(3): 195–202.

34. Khush GS. 2012. Genetically modified crops: the fastest adopted crop technology in the history of modern agriculture. Agric Food Secur [Internet].1(1): 14. Available: http: //agricultureandfoodsecurity. biomedcentral.com/ articles/10.1186/2048- 7010-1-14

35. Garcia-Casal MN, Peña-Rosas JP, Giyose B. 2017. Staple crops biofortified with increased vitamins and minerals: considerations for a public health strategy. Ann N Y Acad Sci.1390(1): 3–13.

36. Paul J-Y, Khanna H, Kleidon J, Hoang P, Geijskes J, Daniells J, *et al.*, 2017. Golden bananas in the field: elevated fruit provitamin A from the expression of a single banana transgene. Plant Biotechnol J. 15(4): 520–32.

37. Naqvi S, Zhu C, Farre G, Ramessar K, Bassie L, Breitenbach J, *et al.*, 2009. Transgenic multivitamin corn through biofortification of endosperm with three vitamins representing three distinct metabolic pathways. Proc Natl Acad Sci U S A.106(19): 7762–7.

38. Singh SP, Gruissem W, Bhullar NK. 2017. Single genetic locus improvement of iron, zinc and β-carotene content in rice grains. Sci Rep. 7(1): 6883.

39. Boncompagni E, Orozco-Arroyo G, Cominelli E, Gangashetty PI, Grando S, Kwaku Zu TT, *et al.*, 2018. Antinutritional factors in pearl millet grains: Phytate and goitrogens content variability and molecular characterization of genes involved in their pathways. Kashkush K, editor. PLoS One [Internet].13(6): e0198394. Available: http://dx.plos.org/10.1371/journal.pone.0198394

40. Zilberman D, Kaplan S, Wesseler J. 2014. The Loss from Underutilizing GM Technologies. J Agrobiotechnology Manag Econ. 2016. 18(3).

41. Wesseler J, Zilberman D. The economic power of the Golden Rice opposition. Environ Dev Econ. 19(06): 724–42.

42. Palmer AC, Siamusantu W, Chileshe J, Schulze KJ, Barffour M, Craft NE, *et al.*, 2016. Provitamin A–biofortified maize increases serum β-carotene, but not retinol, in marginally nourished children: a clusterrandomized trial in rural Zambia. Am J Clin Nutr [Internet].104(1): 181–90.

43. Haskell M, Tanumihardjo SA, Palmer A, Melse-Boonstra A, Talsma E, Burri B. 2017. Effect of regular consumption of provitamin A biofortified staple crops on Vitamin A status in populations in low-income countries. African J Food, Agric Nutr Dev.17(02): 11865–78.

44. Hotz C, Loechl C, de Brauw A, Eozenou P, Gilligan D, Moursi M, *et al.*, 2012. A large-scale intervention to introduce orange sweet potato in rural Mozambique increases vitamin A intakes among children and women. Br J Nutr.108(01): 163–76.

45. Hotz C, Loechl C, Lubowa A, Tumwine JK, Ndeezi G, Nandutu Masawi A, *et al.*, 2012. Introduction of β-Carotene–Rich Orange Sweet Potato in Rural Uganda Resulted in Increased Vitamin A Intakes among Children and Women and Improved Vitamin A Status among Children. J Nutr.142(10): 1871–80.

46. Gannon B, Kaliwile C, Arscott SA, Schmaelzle S, Chileshe J, Kalungwana N, *et al.*, 2014. Biofortified orange maize is as efficacious as a vitamin A supplement in Zambian children even in the presence of high liver reserves of vitamin A: a community-based, randomized placebocontrolled trial. Am J Clin Nutr.100(6): 1541–50.

47. Talsma E, Brouwer I, Verhoef H, Mbera G, Mwangi A, Demir A, *et al.*, 2016. Nutrition News for Africa Biofortified yellow cassava and vitamin A status of Kenyan children: a randomized controlled trial. Am J Clin Nutr.103(1): 258–67.

48. Ilona P, Bouis HE, Palenberg M, Moursi M, Oparinde A. 2017. Vitamin A cassava in Nigeria: Crop development and delivery. African J Food, Agric Nutr Dev.17(02): 12000–25.

49. De Moura FF, Moursi M, Lubowa A, Ha B, Boy E, Oguntona B, *et al.*, 2015. Cassava Intake and Vitamin A Status among Women and Preschool Children in Akwa-Ibom, Nigeria. Sengupta S, editor. PLoS One. 10(6): e0129436.

50. Fudge J, Mangel N, Gruissem W, Vanderschuren H, Fitzpatrick TB. 2017. Rationalising vitamin B6 biofortification in crop plants. Curr Opin Biotechnol. 44: 130–7.

51. Zhang P, Ma Q, Naconsie M, Wu X, Zhou W, Yang J. 2017. Advances in genetic modification of cassava. Achiev Sustain Cultiv.

52. Strobbe S, Van Der Straeten D. Toward 2018. Eradication of B-Vitamin Deficiencies: Considerations for Crop Biofortification. Front Plant Sci [Internet]. 9: 443.

53. Mène-Saffrané L, Pellaud S. 2017. Current strategies for vitamin E biofortification of crops. Curr Opin Biotechnol. 44: 189–97.

54. Haas JD, Luna S V, Lung'aho MG, Wenger MJ, Murray-Kolb LE, Beebe S, *et al.*, 2016. Consuming Iron Biofortified Beans Increases Iron Status in Rwandan Women after 128 Days in a Randomized Controlled Feeding Trial. J Nutr. 146(8): 1586–92.

55. Murray-Kolb LE, Wenger MJ, Scott SP, Rhoten SE, Lung'aho MG, Haas JD. 2017. Consumption of Iron-Biofortified Beans Positively Affects Cognitive Performance in 18- to 27-Year-Old Rwandan Female College Students in an 18-Week Randomized Controlled Efficacy Trial. J Nutr [Internet].147(11): jn255356.

56. De Moura FF, Palmer AC, Finkelstein JL, Haas JD, Murray-Kolb LE, Wenger MJ, *et al.*, 2014. Are Biofortified Staple Food Crops Improving Vitamin A and Iron Status in Women and Children? New Evidence from Efficacy Trials. Adv Nutr.5(5): 568–70.

57. Bouis HE, Saltzman A. 2017. Improving nutrition through biofortification: A review of evidence from HarvestPlus, 2003 through 2016. Glob Food Sec.12: 49–58.

58. Jones KM, de Brauw A. 2015. Using Agriculture to Improve Child Health: Promoting Orange Sweet Potatoes Reduces Diarrhea. World Dev [Internet]. 74: 15–24.

59. Okello JJ, Lagerkvist CJ, Muoki P, Heck S, Prain G. 2017. Does Information on Food Production Technology Affect Consumers' Acceptance of Biofortified Foods? Evidence from a Field Experiment in Kenya. J Agric Food Inf.1– 18.

60. Covic N, Low J, MacKenzie A, Ball A. 2017. Advocacy for biofortification: Building stakeholder support, integration into regional and national policies, and sustaining momentum. African J Food, Agric Nutr Dev [Internet].17(02): 12116–29.

61. De Groote H, Kimenju SC, Morawetz UB. 2011. Estimating consumer willingness to pay for food quality with experimental auctions: the case of yellow versus fortified maize meal in Kenya. Agric Econ. 42(1): 1–16.

62. Beyene G, Solomon FR, Chauhan RD, Gaitán-Solis E, Narayanan N, Gehan J, *et al.*, 2018. Provitamin A biofortification of cassava enhances shelf life but

reduces dry matter content of storage roots due to altered carbon partitioning into starch. Plant Biotechnol J [Internet].16(6): 1186– 200.

63. Tanumihardjo SA, Ball A-M, Kaliwile C, Pixley KV. 2017. The research and implementation continuum of biofortified sweet potato and maize in Africa. Ann N Y Acad Sci [Internet].1390(1): 88–103.

64. Ekpa O, Palacios-Rojas N, Kruseman G, Fogliano V, Linnemann AR. 2018. Sub-Saharan African maize-based foods: Technological perspectives to increase the food and nutrition security impacts of maize breeding programmes. Glob Food Sec.17: 48–56.

( Chapter 13 )

# Stress Resilient Crop Production for Sustainable Agriculture

**R.H. Kanth, O.A. Wani and S. Fayaz**

*Faculty of Agriculture, SKUAST-K, Wadura*

India's agricultural growth has been phenomenal over last four decades as the country moved from severe food crisis before 1960's to self-sufficiency and surplus food grain production. Most of this increase in agricultural output could be attributable to green revolution under irrigated environments. However, increase in the area under cultivation as it occurred in the early stages of green revolution, is no longer feasible and in fact there is a decline in the last two decades owing to urbanization and rapid industrialization. Currently the country is facing a challenge of shrinking natural resource base to meet the demands of ever increasing population. Hence, intensification and diversification of agriculture through enhanced productivity and resource use efficiency has to be the main focus as competition for land and water are increasing from non-farm sectors. Further in recent years climate change and its variability are emerging as major challenges to Indian agriculture. Although, the impacts of climate change are global, countries like India are more vulnerable in view of the high population depending on agriculture, excessive pressure on natural resources and poor coping capabilities. In view of these, immediate thrust is needed on enhanced production with reduced natural resources under a variable climate. Agriculture in India is predominantly rainfed with nearly 56 per cent of the total cultivated area and contributes about 40 per cent of the country's food production (Venkateswarlu and Prasad, 2012, Srinivasa Rao *et al.* 2017). In addition to the temporal variation of the environment, there is also a large spatial variation in the rainfed belt. Feeding the ever- increasing population remains an uphill task with the dwindling natural resource base along with to enhance risks due to climatic adversities.

Identifying stress tolerant varieties for different agro-ecologies of the country is essential to sustain and accelerate the productivity to meet the increasing demand of food. Tolerant crop varieties with consistently higher yields under deficit and excessive rainfall and other abiotic stresses, such as temperature extremes, salinity *etc.* is of paramount importance. Further, integrated and efficient agronomic management strategies including optimal time of sowing, nutrient and pest management strategies contribute immensely for realizing the maximum genetic potential.

## Climate Change and Agriculture

Indian Agriculture is projected to be affected by the climate change and adaptation is indispensable for developing resilient agricultural system. For Indian region, the AR5- WGII (IPCC 2014) report projects an increase in frequency of extreme temperatures, rainfall, heat wave, flood and drought events and skewed monsoon years. Further, it projects an increased risk of drought-related water and food shortage if agriculture is not adapted to changing climates. The projections of global climate change include altered average temperatures, rainfall, increased extreme events, enhanced atmospheric carbon dioxide, ground-level ozone concentrations and rise in sea level leading to inundation of coastal areas. The extreme events such as heat and cold waves, flooding, hail storms, cyclones are well known to adversely affect the agricultural sector. In recent past it is more evident, as one or the other part in the country is affected by droughts, excessive rains, floods, cyclones, frost, heat wave and other climatic events. The global and regional impacts of projected climate change are expected to be significant on agriculture, water resources, natural ecosystems and food security (IPCC 2014). Although, the impacts of climate change are being experienced globally, countries like India are more vulnerable in view of the large population depending on agriculture. Small and marginal farmers especially of rainfed areas are likely to be more vulnerable to the risks of climate change due to harsher environments and poor coping abilities.

The major agents of climate change have been ascribed to the increased levels of greenhouse gases (GHGs) beyond their natural limits due to the uncontrolled activities such as burning of fossil fuels, increased use of refrigerants *etc.* Agriculture sector also contributes to climate change through emissions of GHGs as well as its expansion to non-agricultural land.

## Impact of Weather Aberrations on Crops

Climate change is projected to have negative effects on irrigated crop yields across regions in India both due to temperature rise and changes in water availability. While rainfed agriculture is primarily impacted due to rainfall variability and reduction in number of rainy days (Venkateswarlu and Shankar, 2012). Shifts in seasons, increase in temperatures and changes in rainfall pattern are already visible. In view of these, the crops may encounter major abiotic stresses like drought, flood, heat and cold during its life cycle, resulting in substantial yield losses. The impacts of these stresses may vary with region, crop and cropping systems, soils and management practices. The yield reductions are likely to be caused by shortening of growing period, negative impacts on reproduction, grain filling and extremes in water availability and temperatures at critical growth stages. The negative impacts due to terminal heat in the month of January/February, increased water stress and reduction in number of rainy days on yield of wheat and paddy are already being felt (Rao and Bapuji Rao, 2013). Weather aberrations impact the crop yields both directly and indirectly. The direct effects are mainly due to change in crop duration and impact reproductive processes such as pollination and fertilization. While the indirect effects are largely due to changes in water availability, altered pest, disease and weed dynamics. The impacts of climate change are mostly crop specific, as the model outputs reveal that the yields of wheat, rice and maize will decrease while it could be neutral or positive with groundnut, soybean and chickpea (Aggarwal, 2008; Aggarwal, 2018).

Rainfed crops are more vulnerable to climate change because of the limited options for coping with variability of rainfall and temperature. This will result in shift in sowing time and shorter growing season, which may necessitate effective adjustment in sowing and harvesting dates. Frequent and more intense extreme events may become the norm of the day for common farming community (IPCC 2014). About 56 per cent of the cultivated area in India is still rainfed, despite the substantial progress in bringing more areas under irrigation in post-independence period. It is estimated that even after achieving the full irrigation potential, >50 per cent of the net cultivated area will remain dependent on rainfall (Sharma, 2011, Srinivasa Rao *et al.*, 2015). There are various factors that can be affected due to weather aberration, especially in areas under rainfed where more than 80 per cent farmers are small and marginal (with < 1 ha of land).

Long dry spells of early season, mid-season and terminal drought affect production adversely (Sharma *et al.*, 2006). Early season drought generally occurs either due to delayed onset of monsoon or due to prolonged dry spell soon after the onset of the rainy season. Mid- season drought occurs due to inadequate soil moisture availability between two successive rainfall events during the crop growth period. Late season or terminal drought occurs as a result of early withdrawal of monsoon rains. Water stress at any stage of crop growth cycle will adversely impact the productivity, while terminal droughts are more critical as the reproductive stage is highly sensitive. Water stress, which is mostly associated with an increase in ambient temperatures, results in forced maturity. Drought and heat stress at terminal stage of crop are high in the northern, western and central India, resulting in high yield loss in case of major food crops such as wheat. This necessitates the real time implementation of contingency plans to overcome the adverse impacts of weather aberration in agriculture. In recent past, this has been demonstrated and realized under actual field conditions through successful adoption of flood tolerant rice varieties like Swarna Sub1 in coastal districts while, drought tolerant and high yielding groundnut variety Narayani in the district of Anantapur, Andhra Pradesh. Similarly, in case of wheat, Lok-1 is one the most successful variety grown under heat stress in states of Gujarat, Madhya Pradesh and other areas where crop is exposed to terminal heat stress during grain filling stage and maturity.

# Strategies for Coping with Weather Aberrations in Indian Agriculture

Adaptation and mitigation strategies including use of climate resilient crops and varieties for different regions are most essential for agriculture to successfully cope with climate variability. Improved agricultural practices for diverse agro-ecological regions in India have potential to enhance climate change adaptation (Venkateswarlu *et al.*, 2011; Aggarwal *et al.*, 2018). Natural resource management practices for adverse climatic conditions aid in enhancing resilience under variable climate and extreme events. Major strategies of adaptation to climate change include water saving technologies such as in-situ and ex-situ moisture conservation, water harvesting for supplemental irrigation, residue incorporation (to avoid it's burning), growing tolerant crop varieties, conservation agriculture, site specific nutrient management practices *etc*. In this process developing and promoting crop varieties with tolerance to abiotic stresses like drought, heat, submergence for the target vulnerable areas is of great significance. Indeed, climate resilient crop varieties play a crucial role for coping with climate variability in agriculture. Further, strengthening institutional interventions in promoting collective action and build resilience among communities through improved varieties along with appropriate national resource management technologies will go a long way in sustaining agricultural production system in the country.

Plant's response to abiotic stresses is crop and variety specific. For example, in case of pigeonpea, higher temperatures will shorten crop duration so that it matures when the wet season is still active, while, sorghum experiences shortening of the vegetative phase relative to the grain-filling phase resulting in increased harvest index. Understanding of photoperiod sensitivity, genetic variation for transpiration efficiency will help in identifying short duration high yielding varieties that escape the terminal drought as well as other impending abiotic and biotic stresses. Indian National Agricultural Research System (NARS) including various ICAR institutes and state agricultural universities are making concerted efforts over the years for developing improved varieties of different crops with enhanced tolerance to multiple abiotic stresses. These varieties could be utilized by the farming communities in the event of extreme weather situations. Climate resilient crop varieties along with suitable adaptation and mitigation strategies will help to overcome the adverse impact of climate change by lowering the yield losses under stress conditions.

## Climate Resilient Crop Varieties for Different Abiotic Stresses

The development and identification of climate resilient varieties with enhanced tolerance to heat, drought, flooding, chilling and salinity stresses are essential to sustain and improve crop yields and to cope with the challenges of climate change. It is essential to enhance the productivity and profitability of farming community by minimizing risk in agriculture in order to improve the livelihoods of millions of people dependent on agriculture. While, abiotic stresses such drought, heat or cold may trigger a series of responses in plants that include changes in gene expression, signal transduction pathways, metabolic and molecular mechanisms as well as cumulative manifestations of these in terms of source and sink relations for adaptation. The major biotic and abiotic stresses limiting crop productivity are given in the following figure. Among various abiotic stresses, drought, heat, salinity, cold and flooding are the major factors that adversely affect plant growth and productivity (Maheswari *et al.*, 2012).

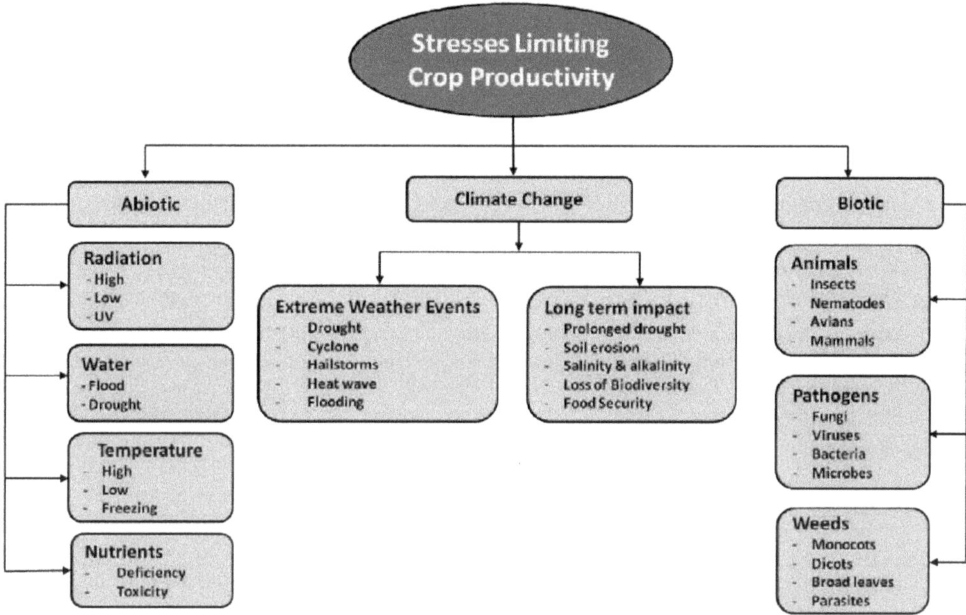


All these adverse environmental conditions have potential to drastically reduce yields in warmer regions. To develop stress tolerant varieties, it is essential to identify the traits that maintain and promote the growth and development of plants during the stress period (Shanker *et al.*, 2014; Maheswari *et al.*, 2016, Maheswari 2017). The tolerance to a particular stress is related to the plant's ability to withstand adverse conditions, survive and reproduce successfully. The tolerance to abiotic stresses is manifested in terms of the ability to cope with resource limitation under stress as well as the ability to recover along with high production potential when stress is relieved. In several crops, genetic control of both stress tolerance and resource-use efficiency is quantitatively inherited involving many loci distributed in different regions of the genome (Wu *et al.*, 2011). Quantifying and understanding the genetic relationship between these two is the key to improve productivity of crops by developing climate resilient varieties. Several crop improvement programs are focused on improving productivity with tolerance to various abiotic stresses *viz.*, drought, heat, cold, salinity, flooding *etc.* The availability of climate resilient crop varieties along with sufficient quantities of quality seeds of these need to be available to the farmers for sustaining the production system and meeting the increasing demand of food grains. Farmers require varieties that produce a satisfactory yield when subjected to stress conditions but also have a high productivity potential under favourable conditions.

# Crop Varieties Suitable for Cultivation under Different Abiotic Stresses

## Drought and Delayed Monsoon

There have been tremendous advances in understanding physiology, biochemistry and molecular genetics of plant responses to different abiotic stresses. Number of adaptive traits have been studied and used for improvement of drought tolerance like early vigour, short duration, osmotic adjustment, leaf senescence, stay green *etc.* Stay green habits in plants, usually refer to tolerance against drought-induced post-flowering senescence. Roots also play an important role in adaptation to drought stress. Various ICAR institutes and state agricultural universities are making concerted efforts to develop high yielding varieties of different crops with enhanced tolerance to delayed monsoon and drought over years which can be utilized by the farming communities. Major food, vegetable and horticultural crop varieties with tolerance to drought stress, and delayed monsoon and varieties with short duration released by various Institutes/Universities. are given in the following tables.

## Cold Stress

Major food crops such as maize (*Zea mays*) and rice (*Oryza sativa*) are highly sensitive to low temperatures. The growth of these crops is severely affected in terms of growth and development by temperatures below 10°C resulting in considerable yield loss or even crop failure. When the temperature decreases to less than 5°C for more than three consecutive days it is considered as cold wave/stress in areas where normal temperature remains 10°C or above, while in areas where normal temperature is below 10°C, if temperature goes below 3°C for more than three days it is considered as cold wave (Venkateswarlu *et al.*, 2011). Many plants, especially those, which are native to warm habitat, exhibit symptoms of injury when subjected to low non-freezing temperatures. Various symptoms in response to cold/chilling stress include reduction of leaf expansion, wilting, chlorosis and necrosis. In chilling stress, primary injury is the initial rapid response that causes a dysfunction in the plant, but is readily reversible if the temperature is raised to non-chilling conditions (Kratsch and Wise 2000).

## Salinity Stress

Salinity is another important abiotic stress limiting crop production worldwide. Increased salinization of arable lands is expected to have devastating effects on agricultural production in many countries including India. High salinity causes both hyper-ionic and hyper- osmotic stress and can lead to plant death (Munns and Tester 2008). Salinity in a given land area depends on the amount of evaporation in relation to the amount of precipitation leading to increase in salt concentration. Intrusion of sea water also is another major cause for increase in salinity. Sodicity is a secondary result of salinity in clay soils, where leaching washes soluble salts into the subsoil, while sodium is left bound to the negative charges of the clay (Wang *et al.*, 2003). Agricultural lands that have been heavily irrigated are becoming highly saline, while in drier areas there is extensive water loss through a combination of both evaporation as well as transpiration. High salt concentration (Na+) deposited in the soil alter the basic texture of the soil resulting in decreased soil porosity, reduced soil aeration and water conductance. While, several crops are sensitive to salinity, rice can thrive relatively better on salt-affected soils as standing water helps in leaching salts from topsoil. A number of mapping studies have been attempted to identify QTLs located on different chromosomes for salinity tolerance in rice. A major QTL designated as 'SALTOL' was mapped on chromosome 1 which accounts for more than 40 per cent of the variation in salt uptake. Several lines containing SALTOL QTL were developed through marker assisted breeding.

## Flooding or Submergence

Water logging, also called as flood/submergence, anoxia, hypoxia *etc.*, is one of the major harmful abiotic stresses limiting crop yields. Generally, flooding in the field can be either water logging in which root and some portion of the shoot under water are complete submergence or where the whole plant is under water. Lack of oxygen supply is the main cause of damage in water logging conditions, because of which plant shifts its metabolism from aerobic to anaerobic mode. Aerenchyma formation, greater activity of glycolate pathway, involvement of anti-oxidative metabolism is some of the adaptive mechanisms to cope with flooding tolerance. Ethylene is associated with induction of genes related to the adaptive mechanism of tolerance (Alamgir and Uddin 2011).

## Strategies for Ensuring Access to Resilient Crop Varieties

The changing climate is a major impediment in sustaining agricultural productivity especially to small and marginal farming communities, where the event of loss of even a single crop can lead to starvation or malnutrition of the family. In case of early season stress, the loss of standing crop at initial stages could be compensated by re-sowing immediately. However, there may not be sufficient seed left with the farmer for re-sowing. Moreover, it would be difficult for the public and private seed sectors to meet the demand of seeds to farmers specially those residing farther from seed source. To overcome such a situation, development of community seed banks may be useful as a contingency measure to meet local seed demand. Ensuring the supply of quality seeds of climate resilient crop varieties along with adoption of suitable adaptation and mitigation technologies is essential to make agriculture more sustainable in the era of climate change. Rainfed agriculture which is more vulnerable to climate change, needs a robust decentralized seed system that is able to provide quality seed of diverse crops and varieties at affordable prices at right time to improve productivity. This will aid in buffering contingencies of climate risks such as repeat sowing in case of crop failure. There is also need to ensure conservation of the local agro biodiversity which has inbuilt tolerance to various stresses.

It is necessary to ensure quality of farm saved seeds for enhancing crop productivity, as in our country, farmers often use these for subsequent crops. The need to replenish diversity in agricultural systems will encourage farming communities to build up community seed banks. This facilitates the revival and distribution of traditional and stress-tolerant crops and varieties. Various aspects of seed production, seed distribution and storage condition have to be improved and strengthened at the farmers' level under a community-based seed system by making a cluster of villages or block as a seed village to cater the quality seed requirement of specified area. The implementing agencies which can play a pivotal role in further strengthening the seed village concept are State Departments of Agriculture, State Agriculture Universities, Krishi Vigyan Kendras, State Seeds Corporation, National Seeds Corporation State Seeds Certification Agencies *etc.* in a coordinated effort.

Seed village and seed banks concept is being implemented successfully in different parts of the country which can further be upscaled to cope with climate variability such as, community seed banks for flood tolerant rice varieties of Bihar and Bengal, saline-resistant rice varieties of Orissa (Wajih 2008).

## Conclusions

The need for stress tolerant varieties has become very important in the present context of climate change apart from various adaptation and mitigation strategies to feed the ever-increasing population in the country. Concerted efforts of the National Agricultural Research System (NARS) during the last few decades resulted in development of stress tolerant varieties in several crops. These efforts are further being strengthened to develop varieties tolerant to various abiotic stresses individually as well as those with multiple stress tolerance. These tolerant varieties can play an important role in coping with climate variability as well as enhancing the productivity. Location specific conservation techniques, water harvesting and efficient management of water resources and other adaptation strategies as well as enabling policies on crop insurance, along with robust early warning system and weather-based advisories will further facilitate enhancing the resilience of Indian agriculture to climate change and climate variability.

## Way Forward

In the context of climate resilient agriculture, stress tolerant crop varieties are one the most important resources in enhancing resilience of the farming community and to efficiently cope with climate variability. Improved and tolerant climate resilient varieties along with the proper management practices can enhance the coping ability through risk reduction in vulnerable environments. Ensuring seed availability of the resilient varieties in various crops at the appropriate time to the farmers is an important challenge to be addressed immediately. Issues related to managing trade-off between risk and expected returns in vulnerable areas to weather aberrations including drought, flood, heat and cold waves *etc.*, also need urgent attention. Participatory approach to consolidate the involvement of village institutions will go a long way in ensuring the seed availability of resilient varieties locally. These resilient varieties need to be included in contingency plan implementation in synergy with national programmes such as RKVY, NMSA *etc.* Another important dimension of utilizing the present information on climate resilient varieties is that, these could be utilized as potential genetic resources for further advancement using tools of both conventional as well as marker assisted selection and other cutting-edge science tools.

# REFERENCES

Aggarwal, P. K. 2008. Global climate change and Indian agriculture: Impacts, adaptation and mitigation. Ind. J. Agri. Sci. 78: 10-16.

Aggarwal, P. K., Jarvis, A., Campbell, B. M., Zougmoré, R. B., Khatri-Chhetri, AVermeulen,., S. J., Loboguerrero, A., Sebastian, L. S., Kinyangi, J., Bonilla-Findji, O., Radeny, M., Recha, J., Martinez-Baron, D., Ramirez-Villegas, J., Huyer, S., Thornton, P., Wollenberg, E., Hansen, J., Alvarez-Toro, P., Aguilar-Ariza, A., Arango-Londoño, D., Patiño-Bravo, V., Rivera, O., Ouedraogo, M. and Tan Yen, B. 2018. The climate-smart village approach: framework of an integrative strategy for scaling up adaptation options in agriculture. Ecol. and Soc. 23: 14. https: //doi.org/10.5751/ES-09844-230114

Alamgir H. and Uddin, S. N. 2011. Mechanisms of waterlogging tolerance in wheat: morphological and metabolic adaptations under hypoxia or anoxia. Aus. J. Crop Sci. 5: 1094–1110.

Auffhammer, M., Ramanathan, V. and Vincent, J. R. 2012. Climate change, the monsoon, and rice yield in India. Climatic Change, 111: 411-424.

Goswami, B.N. 2006. Increasing trend of extreme rain events and possibility of extremes of seasonal mean Indian monsoon in a warming world (http: // saarc-sdmc.nic.in/pdf/workshops/ kathmandu/pres16.pdf).

IPCC, 2014: Summary for policymakers. In: Climate Change 2014: Impacts, Adaptation, and Vulnerability. Part A: Global and Sectoral Aspects. Contribution of Working Group II to the Fifth Assessment Report of the Intergovernmental Panel on Climate Change [Field, C.B., V.R. Barros, D.J. Dokken, K.J. Mach, M.D. Mastrandrea, T.E. Bilir, M. Chatterjee, K.L. Ebi, Y.O. Estrada, R.C. Genova, B. Girma, E.S. Kissel, A.N. Levy, S. MacCracken, P.R. Mastrandrea, and L.L. White (eds.)]. Cambridge University Press, Cambridge, United Kingdom and New York, NY, USA, pp. 1-32.

Kratsch, H.A. and Wise, R.R. 2000. The ultrastucture of chilling stress. Pl. Cell Env. 23: 337-350.

Kulkarni, A., Deshpande, N., Kothawale, D.R., Sabade, S.S., Rama Rao, M.V.S., Savin, T.P.,Patwardhan, S., Mujumdar, M., and Krishnan, R. 2017. Observed climate variability and change over India. In : Climate change over India – An Interim report (Eds. R. Krishnan, J. Sanjay). ESSO-IITM, GoI, http: //cccr.tropmet.res. in/home/docs /cccr/climatechangereport

Maheswari, M. 2017. Enhancing tolerance to climatic stresses in rainfed crops: The road ahead. p. 105-111. In: Agriculture under Climate Change: Threats, Strategies and Policies. Eds. Belavadi, V.V., Nataraja K.N. and Gangadharappa, N.R. Allied Publishers. ISBN: 978-93-85926-37-2.

Maheswari, M., Yadav, S.K., Shanker, A. K., Anil Kumar, M. and Venkateswarlu, B. 2012. Overview of Plant stresses: Mechanisms, Adaptations and Research Pursuit. In: Crop stress and its management: Perspectives and strategies. Eds. Venkateswarlu, B., Arun Shanker, K. and Maheswari, M. Springer Dordrecht, Heidelberg, London, New York. p1-18.

Maheswari M, Vijaya Lakshmi, Varalaxmi Y, Sarkar B, Yadav S K, Jainender Singh, Seshu Babu G, Ashish Kumar, Sushma A T, Jyothilakshmi N, Vanaja M 2016. Functional mechanisms of drought tolerance in maize through phenotyping and genotyping under well-watered and water stressed conditions. Eur. J. Agron., 79: 43-57. http: //dx.doi.org/10.1016/j.eja.2016.05.008

Malik, N., Bookhagen, B. and Mucha, P. J. 2016. Spatiotemporal patterns and trends of Indian monsoonal rainfall extremes. Geophys. Res. Lett., 43, 1710–1717.

Mukherjee, S., Aadhar, S., Stone, D. and Mishra, V. 2017. Increase in extreme precipitation events under anthropogenic warming in India. Weather and Climate Extremes 30 : 1–9

Munns, R. and Tester, M. 2008. Mechanisms of Salinity Tolerance. Ann. Rev. Plant Biol. 59: 651- 681. https: //doi.org/10.1146/annurev.arplant.59.032607.092911

Rao, V.U.M. and Bapuji Rao, B. 2013. Climate change impact on Indian agriculture – adaptation and mitigation strategies. J. Res. Punjab Agri. Univ. 50: 82-91.

Roxy, M.K., Ghosh, S., Pathak, A., Athulya1, R., Mujumdar, M., Murtugudde, R., Terray, P. and Rajeevan, M. 2017. A threefold rise in widespread extreme rain events over central India. Nature Commun. 8: 708. DOI: 10.1038/s41467-017-00744-9.

Shanker, A.K., Maheswari M., Yadav, S.K., Bhanu, D., Attal N.B. and Venkateswarlu, B. 2014. Drought stress responses in crops. Func. Integr. Gen. 14: 11-22.

Sharma, B. R., Rao, K.V., Vittal, K.P.R. and Amarasinghe. U.A. 2006. Realizing the potential of rainfed agriculture in India. Draft prepared for the IWMI-CPWF Project on Strategic Analyses of India's National River-Linking Project, Colombo, Sri Lanka: International Water Management Institute.

Sharma, K.D. 2011. Rain-fed agriculture could meet the challenges of food security in India. Cur. Sci.100: 1615-16.

Srinivasa Rao, Ch., Lal, R., Prasad, J.V.N.S., Gopinath, K.A., Singh, R., Jakkula, V.S., Sahrawat, K.L., Venkateswarlu, B., Sikka, A.K. and Virmani, S.M. 2015. Potential and Challenges of Rainfed Farming in India. Adv. Agron. 133: 113-181.

Srinivasa Rao, Ch., Rejani R., Rama Rao, C. A., Rao, K. V., Osman, M., Reddy, K.S. Kumar, M. and Kumar, P. 2017. Farm ponds for climate-resilient rainfed agriculture. Current Sci., 112: 471- 477.

Venkateswarlu, B. 2013. Climate Change Scenario in India and its Impact on Agroecosystems. In: Ravindra Chary, G. Srinivasa Rao Ch., Srinivas, K., Maruthi Sankar, G.R., Nagarjuna Kumar,R. and Venkateswarlu, B. 2013. Adaptation and Mitigation Strategies for Climate Resilient Agriculture, Central Research Institute for Dryland Agriculture, ICAR, Hyderabad, India, pp1- 16.

Venkateswarlu, B. and Prasad, J. V. N. S. 2012. Carrying capacity of Indian agriculture: issues related to rainfed agriculture. Cur. Sci. 102: 882-888.

Venkateswarlu, B. and Shanker, A. K. 2012. Dryland agriculture: bringing resilience to crop production under changing climate. In: Crop Stress and its Management: Perspectives and Strategies (pp. 19-44). Springer Netherlands.

Venkateswarlu, B., Singh, A.K., Prasad, Y.G., Ravindra Chary, G, Srinivasa Rao Ch., Rao, K.V., Ramana, D.B.V. and Rao, V.U.M. 2011. District level contingency plans for weather aberration in India. Central Research Institute for Dryland Agriculture, Natural Resource Management Division. ICAR, Hyderabad. P.136.

Wajih, S.A. 2008. Adaptive agriculture in flood affected areas. LEISA Magazine, 24: 24- 25.

Wang, W., Vinocur, B. and Altman, A. 2003. Plant responses to drought, salinity and extreme temperatures: towards genetic engineering for stress tolerance. Planta 218: 1-14.

Wu, Y., Liu, W., Li, X., Li, M., Zhang, D., Hao, Z., Weng, J., Xu, Y., Bai, L., Zhang, S. and Xie, C. 2011. Low nitrogen stress tolerance and nitrogen agronomic efficiency among maize inbreds: comparison of multiple indices and evaluation of genetic variation. Euphytica. 180: 281-290.

( Chapter 14 )

# Redefining Food Security: Why Innovation is Not Enough


**Puja Theil**

*Program Manager, Nelis Global*
*(Next Leaders Initiative for Sustainability, Norway, UK)*


Food is something we as a species have learned to produce and this has changed the way we live. It is in fact the most essential thing we produce. Water is essential too, but we don't really produce it. Technology plays an important role in our lives, but it doesn't determine our chances of living like food does.

Considering the resources, people, land, water and carbon emissions involved in producing food, sustainable food may be the most important step to a sustainable future. Our current food systems are riddled with flaws:

☆ 70 per cent of water, 37 per cent of all land surface and over 25 per cent of the world's workforce is used for food production.

☆ Over 30 per cent of all food produced gets wasted. This does not include food losses at farms or food mass lost due to food processing.

☆ The use of fertilizers and other chemicals has been degrading our land and soil life. According to the UN, we have only 60 years' worth of top soil left, after which we will be farming in dirt.

☆ Fertilizer run-off into streams and rivers creates algal blooms leading to hypoxic conditions deadly for aquatic life and results in dead zones in our oceans.

☆ This inefficient system can't be able to scale up to feed the world's population predicted to grow to about 11 billion people by the end of the century.

A major step in making our food sustainable is to understand the hidden and not so hidden flaws of the food we eat today. There are obvious flaws (food transportation), misunderstood flaws (like GMOs) and more hidden flaws (like modern slavery in our food systems). While there is an army of good-looking innovative solutions tackling big and small challenges, there are less appreciated natural solutions to solve nature's problems. The next decade has been announced as the "International Decade on Ecosystem Restoration" by the UN and we'll look into how this can help food production.

The main goal of this lecture is for us to get to know our food systems a little better - because connecting with a problem is the first step to finding great solutions!

- Agriculture contributes 26% of all GHGs.
  - Climate change and ozone layer depletion

- 37% of the world's surface is used for growing food
  - We lose 20 hectares of land every minute.
  - This loss is compensated for by deforestation.

- 70% of water goes into food production.

- Fertilizer and pesticide run-off result in dead zones and destruction of marine ecosystems.

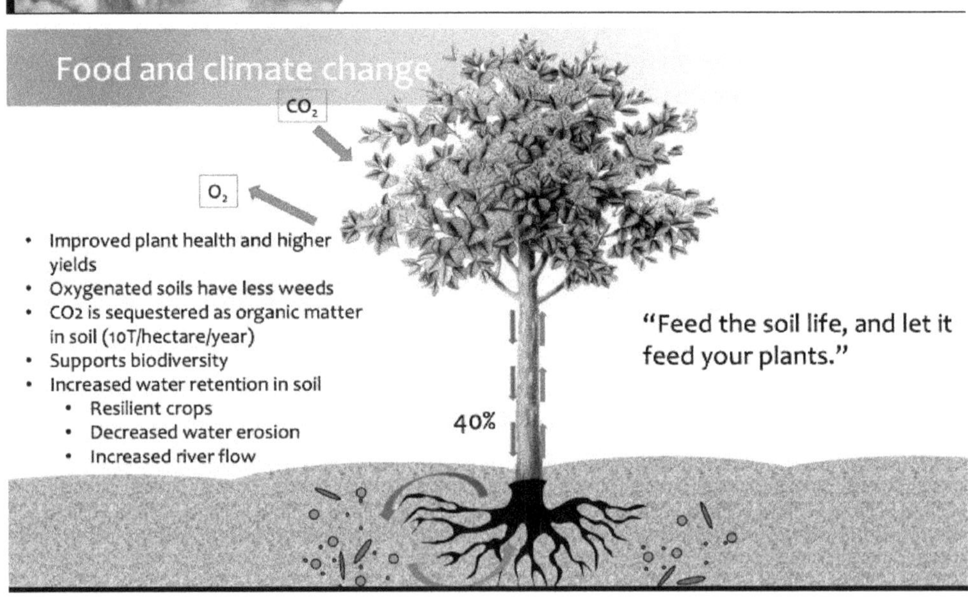

Puja Thiel

## Main Elements of the Lecture

1. Key issues in our "food culture" – At its root, sustainability is a cultural problem, and so are many flaws in our food system. We'll break down some numbers linked to key sustainability issues to be able to think outside the box.

2. Solving climate change with food – Food is not just a sustainability issue, but a potential solution to many sustainability and social issues.

   ☆ Trees are the most efficient technology we have in sequestering carbon.

   ☆ Biomass is a great way of storing away atmospheric carbon (provided it isn't used for fuel).

   ☆ Land degraded by agricultural practices has been restored through regreening efforts in many corners of the world and has helped pull millions of people out of poverty.

# How Countries Can Reduce Child Stunting at Scale: Lessons from Exemplar Countries


**Zulfiqar A. Bhutta**

*Co- Director, Centre for Global Child Health, Hospital for Sick Children, Toronto, Ontario, UK*


## Introduction

The global prevalence of childhood linear growth retardation, commonly measured as stunting, has declined over the last decades, but remains unacceptably high. In most cases, such growth failure is seen in the wake of persistent and widespread poverty, chronic undernutrition, and adverse environmental exposures. The bulk of linear growth faltering seen in children aged <5 y occurs within 1000 d after conception [1,2]. It can start in the womb, in the context of poor maternal nutrition and infection, and extends to the first 2 y of life and sometimes beyond [3]. The 2012 World Health Assembly (WHA) endorsed a 40 per cent reduction in the number of stunted under-5 children by 2025 [4]. This study was funded by a grant to the Centre for Global Child Health from Gates Ventures. The funders had no role in the design, implementation, analysis, and interpretation of the data. Data described in the manuscript, code book, and analytic code will be made available upon request pending application. Publication costs for this supplement were defrayed in part by the payment of page charges. The opinions expressed in this publication are those of the authors and are not attributable to the sponsors or the publisher, Editor, or

*Am J Clin Nutr 2020; 00:1–11. https://doi.org/10.1093/ajcn/nqaa153.*

Editorial Board of *The American Journal of Clinical Nutrition*. Supplemental Figure 1 and Supplemental Tables 1 and 2 are available from the "Supplementary data" link in the online posting of the article and from the same link in the online table of contents at https://academic.oup.com/ajcn/.

A commitment that was reinforced by target 2.2 of the Sustainable Development Goals (SDGs), which explicitly outlines the global pledge to reduce the prevalence of child stunting. Despite these longstanding targets, we are not on track to reach them [5]. Previous *Lancet* series have highlighted the importance of optimizing childhood nutrition for survival[6,7], early child development, and human capital development across the life course [8]. These series also underscore many evidence-based and cost-effective interventions that can improve maternal and child nutritional status and reduce morbidity and mortality [9–11]. Since, new evidence has been generated, including the impacts of lipid-nutrient supplements on stunting, wasting, anemia [12]. Much of the evidence in relation to the potential impact of interventions comes from modest-scale randomized trials and meta-analyses of effect sizes in varying contexts. The 2013 *Lancet* maternal and child nutrition series estimated that scaling up a set of 10 evidence-based "nutrition-specific" interventions to 90 per cent coverage in 34 focus countries with 90 per cent of the global stunting burden could result in a 20.3 per cent (10.2–28.9 per cent) stunting reduction [10]. Econometric analyses done by Smith and Haddad in 2015 [13] and Headey in 2013[14] also sought to quantify the effect of changes in basic and underlying determinants on population-level stunting reductions, and highlighted the importance of agricultural productivity, gender equality, women's education, water and sanitation infrastructure, health services access, and fertility rates—the targets of "nutrition-sensitive" interventions.

The dearth of population-based data on dietary intake has made assessing the contribution of nutrient adequacy, an immediate determinant of linear growth retardation, difficult. In addition, there are limited data on programmatic coverage of high-impact, nutrition-specific interventions. Multiple contributing factors mean that policies, programs, and interventions to drive stunting reduction must involve different sectors and delivery platforms [15]. The determinants and risks associated with concurrent stunting and wasting also merit examination, because linear growth faltering is not the only issue affecting child well-being in vulnerable populations [16–18]. There are a range of overlapping, hierarchical, and potentially intergenerational determinants along the causal chain to poor child growth and stunting that can contribute differently to outcomes in various contexts. The first article of this supplement reviewed the major drivers that have led to national reductions in stunting prevalence, operating at basic, underlying, and immediate levels in >70 countries, the majority of which are low- and middle-income countries (LMICs) [19]. Determinants of particular importance were improved parental literacy rates, household socioeconomic status, water, sanitation, and hygiene (WASH) conditions, health services access, and family planning.

Recent case studies of success have identified these same factors as contributing to change in undernutrition over time in a set of LMICs [20]. These studies also highlighted the role of several factors that enabled policy commitments to nutrition and program strategies to scale up nutrition interventions [20]. In an effort to expand, strengthen, and contribute to the current knowledge base, this article consolidates findings from our series of systematic mixed-methods exemplar case studies with the aim of providing summative guidance on how to accelerate reductions in national stunting prevalence in order to achieve our WHA and SDG targets. Exemplar countries are those that have an outsized reduction in stunting prevalence in a 15–20 year period relative to economic growth (as defined earlier in the "Methods" article of this supplement) [21] and include Peru, Kyrgyz Republic, Nepal, Ethiopia, and Senegal as examples from Latin America, Central Asia, South Asia, East Africa and West Africa, respectively.

Specifically, we

1. Review the current burden, trends, and distribution of child stunting prevalence in exemplar countries;

2. Contrast key basic, underlying, and immediate drivers of child stunting reduction across exemplar countries;

3. Undertake an assessment of policy and programmatic investments across exemplars;

4. Propose a framework for categorizing drivers of change; and

5. Suggest a roadmap for national or subnational efforts to reduce child stunting at scale.

## Methods

Data and evidence for this article were drawn from our country case studies that were undertaken using mixed methods as described in [21]. As detailed in the article, our approach was informed by existing literature and published methodological approaches to study success stories [22]. Initially, a systematic literature review of more than 15 published peer-reviewed and gray literature sources across all LMICs with an expanded search for exemplar countries was undertaken. Studies published between 1990 and 2018 were screened for relevance and synthesized. The aim was to identify and synthesize information on contextual factors, national and subnational interventions, policies, strategies, programs, and initiatives that might have theoretically contributed to stunting reduction in each country over time. Findings from this exercise were used to inform the research process and methodology, and support findings and inferences. The final set of methods included:

1. A series of descriptive analyses of cross-sectional data;

2. A deterministic analysis of quantitative drivers of change in linear growth;

3. Country level stakeholder interviews;

4. A review of policy and program evolution over the time period of study; and

5. An integrative analysis of insights from these 4 core research activities.

1. A series of descriptive analyses were undertaken to explore levels, distributions, and trends in the main outcomes examined, that is, child height-for-age z-scores (HAZ) and under-5 child stunting prevalence (HAZ < –2 SD of the WHO growth reference standards). Using cross-sectional data, we examined stunting prevalence geospatial patterns and inequalities by key dimensions (wealth quintiles, maternal education, urban/rural residence, and child gender). Kernel density plots of child HAZ scores were examined to understand the population HAZ shifts over time. Child age compared with HAZ growth trajectories (Victora curves) were fitted using smoothed polynomial regressions to study stunting risk at birth and how the growth faltering process changes with age.

2. Using household survey datasets (*e.g.*, demographic and health surveys, multiple indicator cluster surveys) and various country-level administrative data sources, we assembled cross-sectional panels into a pseudo-cohort design to study multivariable quantitative drivers of child stunting reduction in each country using Oaxaca–Blinder Decomposition methods and linear mixed-effects regression with a difference-in-difference design.

3. We conducted in-depth interviews with national experts from within and outside the health sector (10–20 participants in each country), in-depth interviews with community-level stakeholders (10–15 participants), and focus group discussions with mothers in communities (2– 4 focus groups in each country with 10–15 participants each) to understand the macro (policy, programs, contextual factors) and micro (individual, household, community) factors that contributed to stunting reduction.

4. A comprehensive review of programs/policies/strategies/laws/legislation adopted from the 1990s (or earlier) to 2018 from within and outside the health sector was undertaken to identify key national investments and efforts that each country adopted to reduce stunting at scale. Detailed methods for this review are included in Paper 2 of this series [21].

5. Results from each research activity were analyzed iteratively and shared with technical experts, research partners, and research leads to develop a strong evidence-based narrative of stunting reduction drivers for exemplar countries. The conceptual framework for improved nutrition proposed by UNICEF [23] and adapted by Black *et al.*[6,7] was used to guide all analyses and inferences (Figure 15.1). Analysis in this article consolidates and synthesizes literature, policy/program, qualitative, and quantitative findings across exemplar countries.

## Results

### Descriptive Trends

Stunting prevalence baselines in exemplar countries generally cluster into 2 groups (Figure 15.2), those that start at 50–60 per cent baselines (Nepal, Ethiopia)

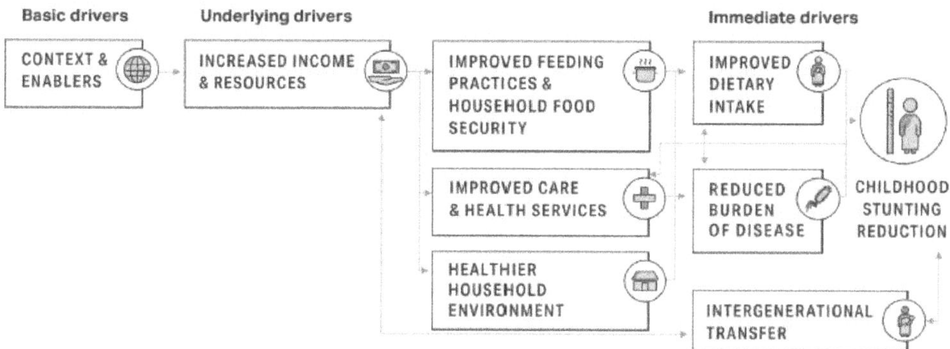

**Figure 15.1: Conceptual Framework for Analyzing Determinants of Child Nutrition (Adapted from references 21, 6 and 7).**

and those with 30–40 per cent baselines (Peru, Kyrgyz Republic, Senegal), and compound annual growth rates range from –2.5 per cent to –5.9 per cent in the ~2000–2016 period.

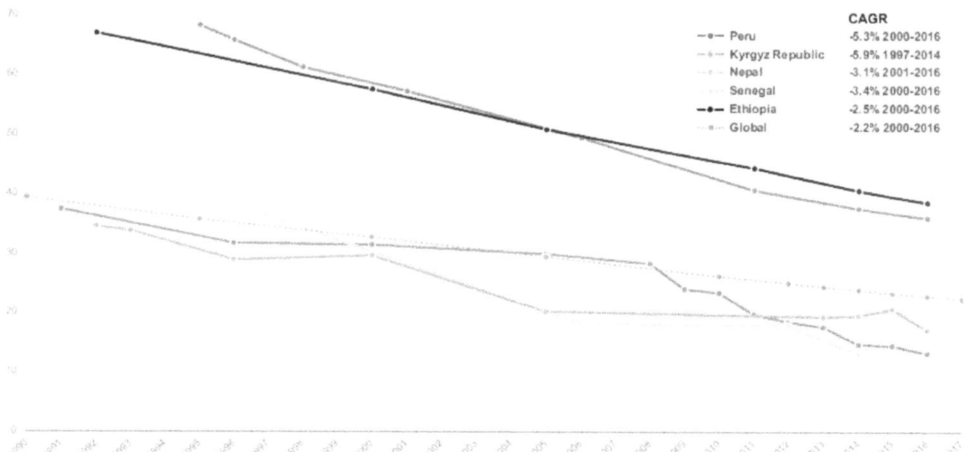

**Figure 15.2: Trends in Child Stunting Prevalence in Exemplar Countries (1990-2016). CAGR: Compound Annual Growth Rate.**

Peru (–5.3 per cent) and Kyrgyz Republic (–5.9 per cent) have almost double the rate of reduction of other countries. Similar progress in stunting reduction across exemplar nations is impressive considering the diverse contextual starting points for each country (Table 15.1). Population size and economic growth vary considerably; for instance, Kyrgyz Republic had only 6.1 million citizens in 2016 whereas Ethiopia had a total population of 102.4 million. All countries saw continued economic growth in the study period (range 3.3–8.2 per cent/y) although their gross domestic product per capita varied vastly in 2016 ($1746 for Ethiopia compared with $13,404 for Peru, purchasing power parity (PPP) measured in current international dollars). Despite differences, across exemplar countries, gains in overall development, employment, literacy, female empowerment, household conditions, health and out-of-pocket spending, health worker availability, and maternal and child health (among other indicators) have been observed (Table 15.1).

Strong stunting reduction was coupled with entire child HAZ distribution shifts (see Supplemental Figure 1 for HAZ kernel density plots) in exemplar countries, suggesting that child growth faltering is reducing at a national level. Examining kurtosis is useful to measure outliers in the data, with higher values indicating fewer outliers (or fewer children in the tails of the HAZ distribution). Kurtosis increased notably over time in Peru and Kyrgyz Republic, suggesting a potential reduction in HAZ inequalities (fewer children in the tails of the distribution) whereas in Senegal, Ethiopia, and Nepal it stayed about the same or even increased. Of course, reduced kurtosis could also indicate better measurement or precision around anthropometry metrics over time, and the difference is difficult to ascertain from HAZ kernel density curves alone (Supplemental Table 15.1). Our equity analysis of stunting prevalence provides further insights into inequalities. Stunting prevalence reduced for all wealth quintiles in Peru (with only the poorest lagging behind) from 2000 to 2016, whereas in Kyrgyz Republic the >20 per cent point gap between the richest and poorest in 1997 became negligible in 2014 (Figure 15.3A). Interestingly, wealth inequalities did not reduce, and in fact, even widened in Nepal, Senegal, and Ethiopia over time. Child stunting prevalence was consistently higher in the least educated mothers, though gaps reduced over time in most exemplar countries (Figure 15.3B). Families living in rural areas had higher stunting prevalence in all countries, though disparities between urban and rural households did reduce in most countries except Ethiopia, where gaps widened from 2000 to 2016 (Figure 15.3C). Our assessment across child gender (Figure 15.3D) revealed minimal differences between boys and girls in general; however, boys were always slightly more stunted than girls in all countries.

Child HAZ compared with age curve analysis (Figure 15.4) suggests that Ethiopia, Nepal, and Kyrgyz Republic have had substantial improvements in length at birth (*i.e.*, mean child HAZ at $y$-intercept increased and became closer to the global reference median of 0). This indicates that newborns became substantially larger at birth over this time, and could be indicative of improvements in maternal nutritional status (for instance, during adolescence and/or prior to conception) and maternal and newborn care. Peru and Senegal did not make progress in reducing the birth disadvantage. Flattening of the Victora curve between 0 and 6 mo was observed in all countries over time, and could be related to better breastfeeding practices and overall environment for the baby. For children aged 6–23 mo, Peru and Senegal had dramatic flattening of the HAZ curve over time (or slowing of growth faltering), which suggests that major stunting drivers in these 2 countries could be related to improvements in disease prevention and management, better dietary practices, and improved household environment. Ethiopia, Nepal, and Kyrgyz Republic made comparatively fewer gains in this age group.

## Table 15.1: Trends in Demographic, Economic and other Contextual Factor for Exemplar Countries[1]

| Indicator | Nepal | | | Peru | | | Senegal | | | Kyrgyz Republic | | | Ethiopia | | |
|---|---|---|---|---|---|---|---|---|---|---|---|---|---|---|---|
| | 2000 | 2016 | CAGR, % | 2000 | 2016 | CAGR, % | 1992 | 2017 | CAGR, % | 2000 | 2016 | CAGR, % | 2000 | 2016 | CAGR, % |
| Total population (millions) | 23.7 | 29.0 | 1.3 | 25.9 | 31.8 | 1.3 | 8.0 | 15.9 | 2.8 | 4.9 | 6.1 | 1.4 | 66.5 | 102.4 | 2.7 |
| Population growth (annual %) | 1.9 | 1.1 | −3.0 | 1.4 | 1.3 | −0.5 | 3.0 | 2.8 | −0.3 | 1.2 | 2.1 | 3.6 | 2.9 | 2.5 | −0.9 |
| Urban population (% of total) | 13.4 | 19.0 | 2.2 | 73.0 | 77.5 | 0.4 | 39.2 | 46.7 | 0.7 | 35.3 | 35.9 | 0.1 | 14.7 | 19.9 | 1.9 |
| Human Development Index | 0.446 | 0.558 (2015) | 1.5 | 0.677 | 0.74 (2015) | 0.6 | 0.37 (1995) | 0.51 | 1.5 | 0.593 | 0.664 (2015) | 0.8 | 0.283 | 0.457 | 3.0 |
| Households with piped water access, % | 35.4 (2001) | 51.4 | 2.5 | 72.3 | 78.9 (2012) | 0.7 | 46.8 | 68.8 | 1.6 | 74.5 (1997) | 82.1 (2012) | 0.7 | 17.6 | 32.0 | 3.8 |
| Open defecation (% engaging in) | 66.9 | 24.0 | −6.2 | 18.3 | 7.1 | −5.8 | 38.9 | 14.9 | −3.8 | 0.1 | 0 | −100.0 | 79.1 | 25.6 | −6.8 |
| GDP per capita, PPP (current international $) | 1215.0 | 2642.0 | 5.0 | 5115.1 | 13,404.2 | 6.2 | 1559.2 | 3556.1 | 3.4 | 1650.8 | 3567.8 | 4.9 | 494.3 | 1746.2 | 8.2 |
| GNI per capita, PPP (current international $) | 1220.0 | 2680.0 | 5.0 | 5000.0 | 12,850.0 | 6.1 | 1530.0 | 3460.0 | 3.3 | 1550.0 | 3380.0 | 5.0 | 490.0 | 1740.0 | 8.2 |
| Poverty headcount ratio at $1.90/d (2011 PPP) (% of population) | 14.5 (2003) | 3.1 (2010) | −19.8 | 16.7 | 3.0 (2015) | −10.8 | 68.4 (1991) | 38.0 (2011) | −2.9 | 42.2 | 2.5 (2015) | −17.2 | 61.2 (1999) | 27.3 (2015) | 6.2 |
| Total unemployment (% of total labour force) | 1.8 | 3.1 | 3.5 | 7.3 | 6.7 | −0.5 | 5.7 (2002) | 6.9 (2015) | 1.6 | 7.5 | 7.2 | −0.3 | 3.7 (1999) | 2.3 (2013) | 6.7 |
| Total health expenditure (% of GDP) | 3.6 | 6.2 (2015) | 3.7 | 4.8 | 5.5 (2014) | 0.9 | 4.6 (2000) | 3.9 (2015) | −1.0 | 4.7 | 6.5 (2014) | 2.3 | 4.4 | 4.0 | −0.6 |
| Public health expenditure (% of GDP) | 1.4 | 2.3 (2014) | 3.6 | 2.7 | 3.3 (2014) | 1.4 | 36.7 (2000)[2] | 31.7 (2015) | −1.0 | 2.1 | 3.6 (2014) | 3.9 | 41.2[2] | 27.6 | −2.5 |
| Out-of-pocket health expenditure (% of total health expenditure) | 55.8 | 60.4 (2014) | 0.6 | 36.4 | 28.6 (2014) | −1.7 | 54.0 (2000) | 44.2 (2015) | −1.3 | 49.8 | 39.4 (2014) | −1.7 | 36.0 | 37.4 | 0.2 |
| Net ODA received (current US$) (millions) | 386.4 | 1065.9 | 6.6 | 401.6 | 331.8 | −1.2 | 662.4 | 736.4 (2016) | 0.7 | 214.7 | 769.0 | 8.3 | 10.3 | 39.8 | 8.8 |
| Age at first marriage (median, women aged 20–49 y) | 17.0 (2001) | 18.1 | 0.4 | 21.4 | 21.6 (2012) | 0.08 | 16.6 | 19.7 (2015) | 1.2 | 20.4 (1997) | 20.6 (2012) | 0.07 | 16.4 | 17.5 | 0.4 |
| Child marriage (% women 20–24 y married by age 18) | 56.1 (2001) | 39.5 | −2.3 | 18.7 | 18.6 (2014) | −0.04 | 47.7 | 31.5 (2016) | −2.6 | 21.2 (1997) | 11.6 (2014) | −3.5 | 49.1 | 41.0 (2011) | −1.2 |
| Fertility rate (average births per woman) | 4.0 | 2.1 | −3.9 | 2.9 | 2.4 (2015) | −1.3 | 6.0 | 4.6 | −1.1 | 2.4 | 3.2 (2015) | 1.9 | 5.9 | 4.6 | −1.5 |
| Births attended by skilled health staff (% of total) | 11.9 | 58.0 | 10.4 | 59.0 | 90.0 (2014) | 3.1 | 47.2 | 68.4 | 1.5 | 98.6 | 98.4 (2014) | −0.01 | 5.6 | 15.5 (2014) | 7.5 |
| Antenatal care ≥4 visits (% of mothers) | 14.3 (2001) | 69.4 (2016) | 11.1 | 68.5 | 94.4 (2012) | 2.7 | 14.0 | 57.0 | 5.8 | No data | 83.6 (2012) | N/A | 10.0 | 32.0 | 7.5 |
| Adolescent fertility rate (births/1000 girls aged 15–19 y) | 113.9 | 62.1 | −3.7 | 65.0 | 48.0 (2015) | −2.0 | 127.0 | 78.0 | −1.9 | 45.9 | 39.2 (2015) | −1.1 | 110.1 | 64.9 | −3.3 |
| Adult literacy rate (% of adults aged ≥15 y) | 48.6 (2001) | 59.6 (2011) | 2.1 | 87.7 (2004) | 94.2 | 0.6 | 39.3 (2002) | 51.9 | 1.9 | no data | 99.2 (2009) | N/A | 35.9 (2004) | 39.0 (2007) | 2.8 |
| Female adult literacy rate (% of females aged ≥15 y) | 34.9 (2001) | 48.8 (2011) | 3.4 | 82.1 (2004) | 91.2 | 0.9 | 29.3 (2002) | 39.8 | 2.1 | no data | 99.0 (2009) | N/A | 22.8 (2004) | 28.9 (2007) | 8.2 |
| Female youth literacy rate (% of females aged 15–24 y) | 60.1 (2001) | 80.2 (2011) | 2.9 | 95.7 (2004) | 98.7 | 0.3 | 41.0 (2002) | 63.5 | 3.0 | no data | 99.8 (2009) | N/A | 38.5 (2004) | 47.0 (2007) | 6.9 |
| Gender Inequality Index (0–1; closer to 1 is higher inequality) | 0.670 | 0.497 (2015) | −2.0 | 0.440 (2005) | 0.385 (2015) | −1.3 | 0.64 (1995) | 0.52 | −0.9 | 0.476 | 0.394 (2015) | −1.3 | 0.6 (2005) | 0.5 | −1.6 |
| Gender Development Index | 0.769 | 0.925 (2015) | 1.2 | 0.921 | 0.949 | 0.2 | 0.781 (1995) | 0.911 | 0.7 | 0.959 | 0.960 | 0.01 | 0.7 | 0.9 | 1.6 |
| Density of physicians (per 1000) | 0.052 (2001) | 0.598 (2014) | 20.7 | 1.166 (1999) | 1.116 (2012) | −0.3 | 0.07 | 0.07 (2016) | 0.0 | 2.8 | 1.9 (2014) | −2.7 | 0.021 | 0.022 (2010) | 0.5 |
| Density of nurses and midwives (per 1000) | 0.469 (2004) | 2.041 (2014) | 15.8 | 0.669 (1999) | 1.299 (2012) | 5.2 | 0.3 (2004) | 0.31 (2016) | 0.3 | 5.5 (2008) | 6.0 (2013) | 1.8 | 0.215 (2003) | 0.236 (2010) | 1.3 |
| Child wasting, % | 11.3 (2001) | 9.7 | −1.0 | 1.1 | 1.0 | −0.6 | 9.0 | 7.2 (2016) | −0.9 | 2.9 | 2.8 | −0.2 | 12.4 | 9.9 | −1.4 |

[1]Sources: World Bank, United Nations Development Programme, Oxford Poverty and Human Development Initiative, Food and Agriculture Organization of the United Nations. CAGR, compound annual growth rate; GDP, gross domestic product; GNI, gross national income; N/A, not applicable; ODA, official development assistance; PPP, purchasing power parity.

[2]This indicator is domestic general government health expenditure (percentage of current health expenditure).

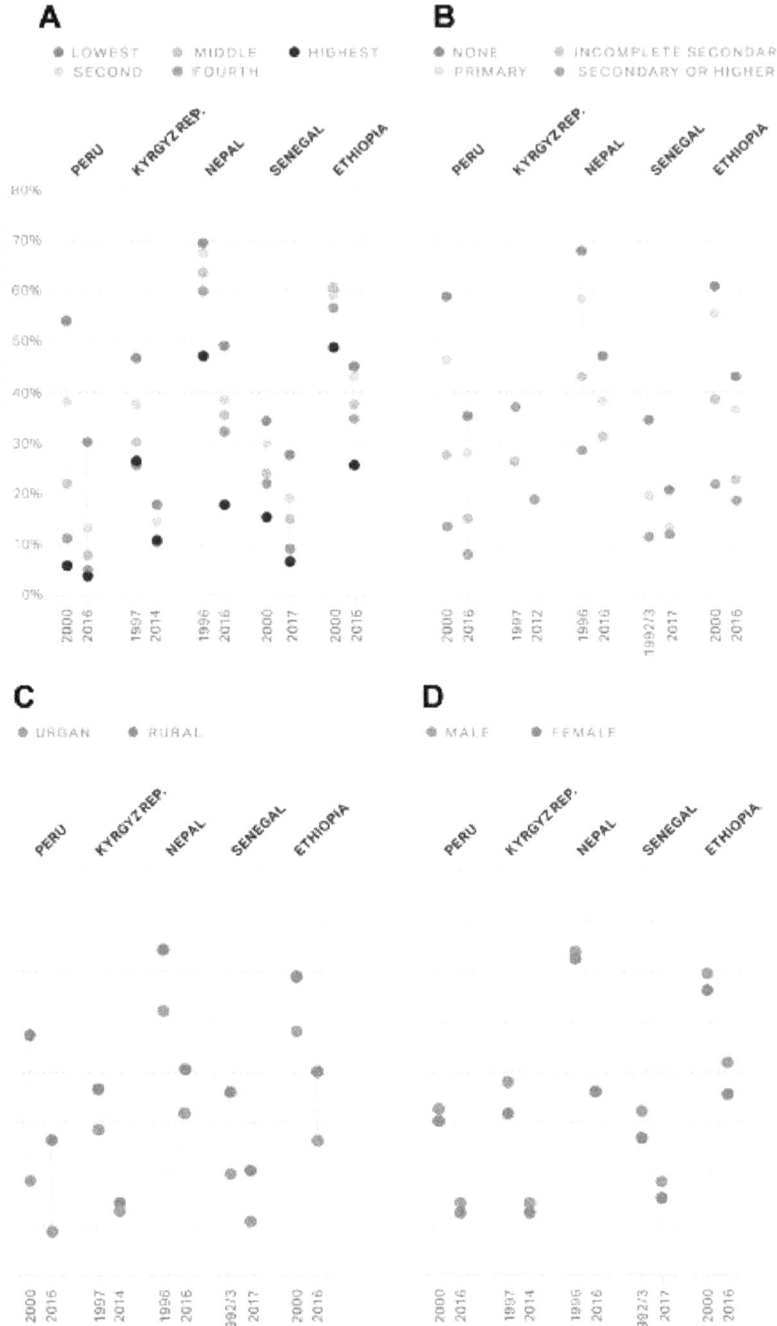

Figure 15.3: (A) Chld stunting prevalence by wealth in exemplar countires; (B) Child stunting prevalence by maternal education in exemplar countries; (C) Child stunting prevalence by residenc in exemplar countries; (D) Child stunting prevalence by chlid sex in exemplar countries.

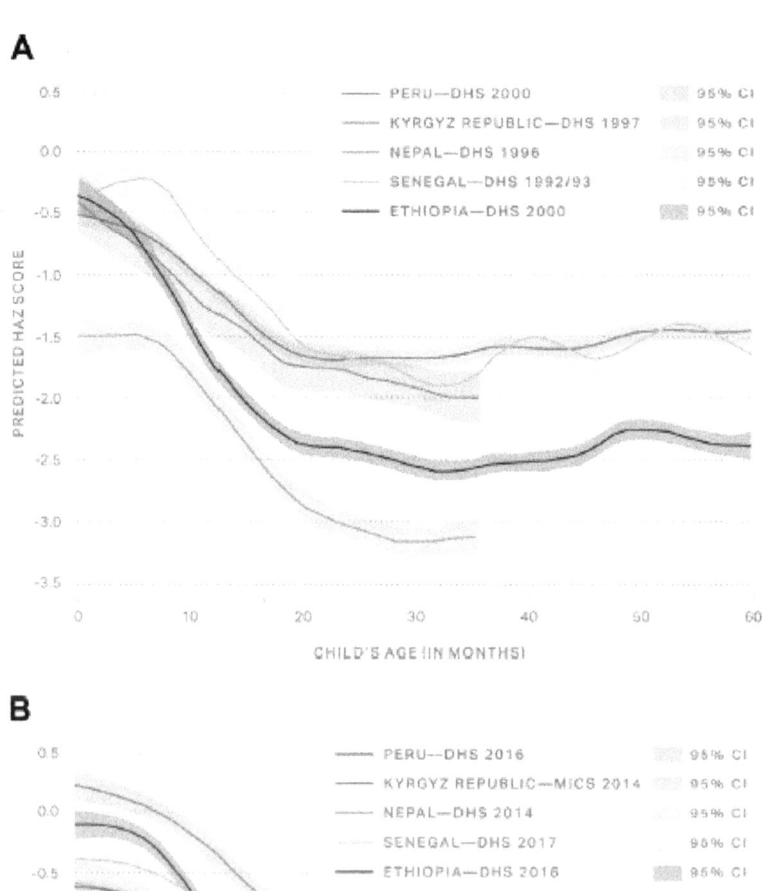


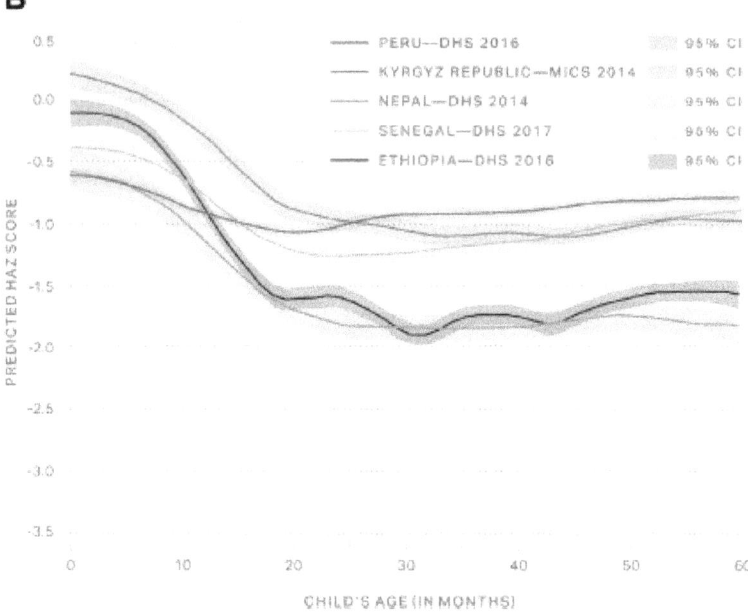


**Figure 15.4: Child Height-for-age Z-score (HAZ) Compared with Age Curves in Exemplar Countries for Earlier Period of Study (A) and Later Period (B); DHS: Demographic and Health Survey; MICS: Multiple Indicator Cluster Survey. Baseline surveys for Nepal and Kyrgyz Republic collected anthropometry for children younger than 3 years only.**

                                                                    

## Determinants of Child HAZ Improvements

When examining quantitative drivers of mean HAZ improvement using multivariable Oaxaca–Blinder decomposition (Figure 15.5), our models could explain 72–100 per cent of mean HAZ improvement across countries. Not withstanding the heterogeneity related to baseline status of health and nutrition indicators and socioeconomic status, the relative contributions of initiatives and factors from within and outside the health sector were evident, and often interventions unrelated to health and nutrition contributed to ~50 per cent of child stunting reduction (of total child HAZ change over time) in each country. The nonhealth sector–related improvements contributed to 36–70 per cent (median 47 per cent) of the observed changes in child HAZ, whereas health- or health care–related changes contributed to 20–64 per cent of change (median 37 per cent). Some of the gaps related to "unexplained" fractions in Kyrgyz Republic, Senegal, and Nepal are likely related to food security/dietary intake or maternal newborn health care or nutrition improvement because proxies for these were lacking in those countries. Notably, improvements in maternal education, maternal and newborn health care, reduction in fertility/reduced interpregnancy intervals, and maternal nutritional status were strong and common contributors to mean HAZ gains across most countries. Some variations in drivers were evident across the countries examined, a finding that largely resulted from differing country contexts and status at baseline. For example,

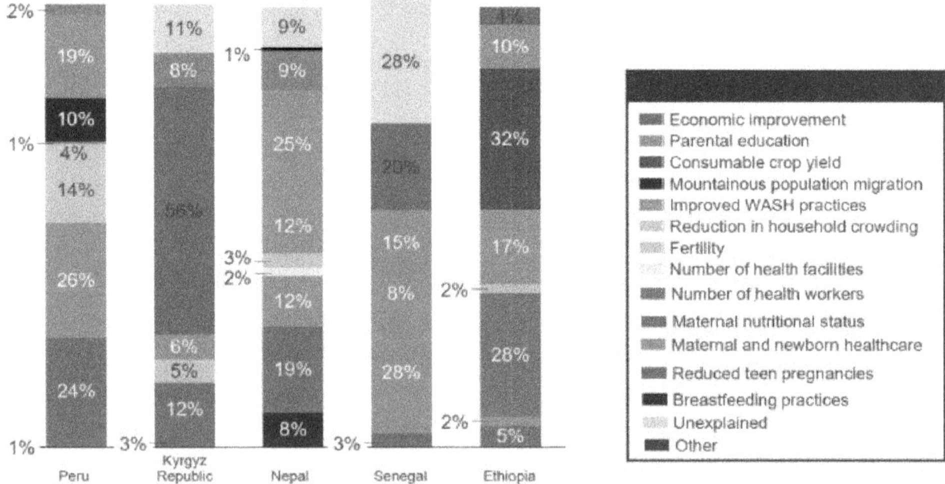


**Figure 15.5: Oaxaca-Blinder HAZ Decomposition Results for Exemplar Countries.**

Note: The Kyrgyz Republic sample is of children aged <3 y. Parental education breakdown: Peru (17.8 per cent maternal, 2.7 per cent paternal), Kyrgyz Republic (5.8 per cent paternal); Nepal (12.2 per cent maternal, 12.5 per cent paternal), Senegal (7.5 per cent maternal, 7.4 per cent paternal), and Ethiopia (5.2 per cent maternal, 5.0 per cent paternal). "Other" category includes child age, gender and region. The following surveys were excluded due to unreliable data: 2014 Kyrgyz Republic MICS and 1996 Nepal DHS. DHS: Demographic and Health Survey; HAZ: Height-for-age-z-score; MICS: Multiple indicator cluster survey; WASH: Water, sanitation and hygiene.

reduced open defecation (from 79 per cent in 2000 to 26 per cent in 2016) contributed to 17 per cent of the change in HAZ in Ethiopia, whereas improved WASH practices did not emerge as a factor in Kyrgyz Republic because of the highly functional water and sanitation systems and practices that were put in place during the Soviet era. In addition, some indicators included in the decomposition analyses were relevant only to 1 country. In Ethiopia, increased consumable crop yield contributed to 32 per cent of the HAZ improvement whereas in Peru, mountainous population migration accounted for 10 per cent.

Although quite different in concept, both indicators could be proxy variables for household food security. Although we did attempt to include determinants at each hierarchical level of the conceptual framework, these analyses are limited by data availability. Regardless, many of the determinants and strategies that contributed to overall HAZ improvement were consistent across countries.

## Country Investments Related to Stunting Reduction

When examining key policy, strategy, and programmatic investments adopted by countries to reduce child stunting, several commonalities were observed (Figure 6). Effective initiatives included health and nutrition interventions addressing immediate determinants that could be grouped into those to improve maternal nutrition and newborn outcomes, promote early and exclusive breastfeeding, and improve complementary feeding practices. In addition, investments in improving reproductive health practices were important for increasing contraceptive use, delaying first pregnancy, and increasing birth spacing. Other sectoral strategies, such as those to improve economic conditions, parental education, and WASH, also played a key role in addressing underlying determinants of child growth. Pivotal to gains were high-level political and donor support as well as sustained financing to improve child health and nutrition overall, and investments in granular data, national capacities for monitoring and decision-making, and capacities for program implementation at scale.

These were most prominent in Peru (especially in the Ministry of Finance) but there were also key enablers of stunting reduction in Nepal, Senegal, and Ethiopia (Figure 15.6).

## A Potential New Framework for Classifying Nutrition-Related Interventions

The 2013 *Lancet* series on maternal and child nutrition was guided by a conceptual framework that outlined the actions needed to achieve optimal child growth and development[7]. This framework is divided into nutrition-specific and nutrition sensitive interventions, programs, and approaches that work to alleviate the immediate and underlying causes of malnutrition, respectively. The framework also outlines the importance of building an enabling environment that will support these strategies through stakeholder buy-in, the generation of rigorous and supportive evidence, advocacy, accountability, coordination, and resource mobilization.

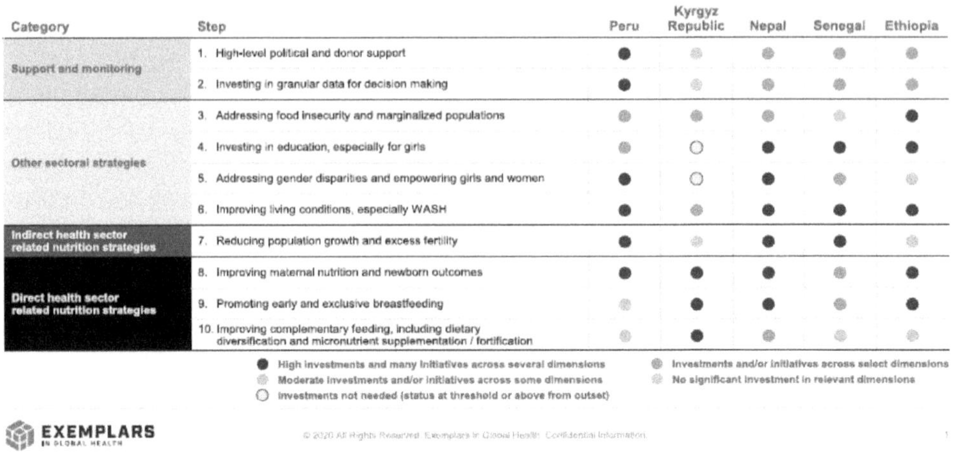

| Category | Step | Peru | Kyrgyz Republic | Nepal | Senegal | Ethiopia |
|---|---|---|---|---|---|---|
| Support and monitoring | 1. High-level political and donor support | ● | ◌ | ◔ | ◔ | ◔ |
| | 2. Investing in granular data for decision making | ● | ◌ | ◔ | ◔ | ◔ |
| Other sectoral strategies | 3. Addressing food insecurity and marginalized populations | ◔ | ◔ | ◔ | ◌ | ● |
| | 4. Investing in education, especially for girls | ◔ | ○ | ● | ● | ● |
| | 5. Addressing gender disparities and empowering girls and women | ● | ○ | ● | ◔ | ◌ |
| | 6. Improving living conditions, especially WASH | ● | ◔ | ● | ● | ● |
| Indirect health sector related nutrition strategies | 7. Reducing population growth and excess fertility | ● | ◌ | ● | ● | ◌ |
| Direct health sector related nutrition strategies | 8. Improving maternal nutrition and newborn outcomes | ● | ● | ● | ◔ | ● |
| | 9. Promoting early and exclusive breastfeeding | ◌ | ● | ● | ◌ | ● |
| | 10. Improving complementary feeding, including dietary diversification and micronutrient supplementation / fortification | ◌ | ● | ◔ | ◌ | ◌ |

Legend:
● High investments and many initiatives across several dimensions
◌ Moderate investments and/or initiatives across some dimensions
○ Investments not needed (status at threshold or above from outset)
◔ Investments and/or initiatives across select dimensions
◌ No significant investment in relevant dimensions



**Figure 15.6: Policy, Strategy and Programmatic Investments Related to Stunting Reducation in Exemplar Countries. WASH: Water, sanitation and hygiene.**

Using this approach, nutrition sensitive programs, such as those relating to agriculture and food security are being leveraged to serve as delivery platforms for nutrition-specific interventions. However, as the evidence base on the impact of nutrition-specific and -sensitive strategies has expanded, and our understanding of what it takes to govern and scale these multisectoral approaches has advanced, the need for further refinement of this framework has become clear. Building on past efforts (24, 25), we are now proposing that direct or indirect nutrition

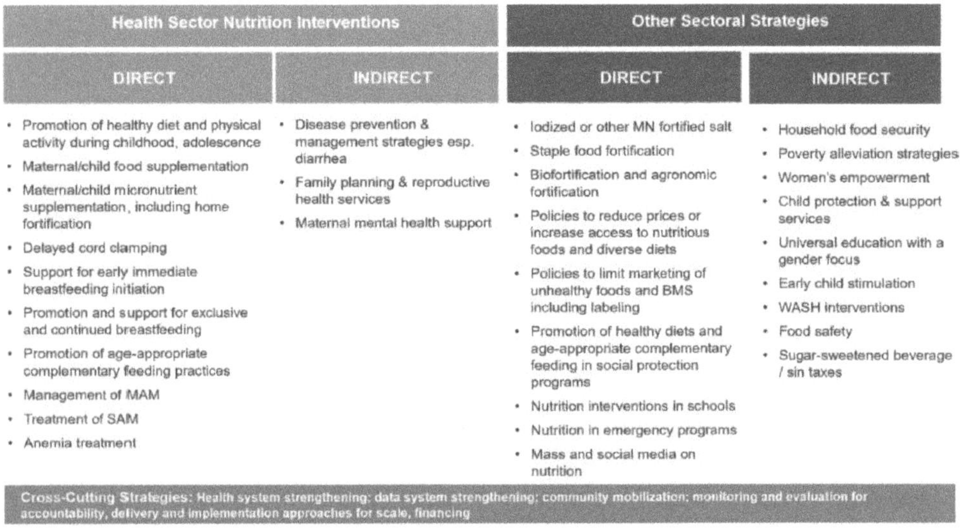

| Health Sector Nutrition Interventions | | Other Sectoral Strategies | |
|---|---|---|---|
| **DIRECT** | **INDIRECT** | **DIRECT** | **INDIRECT** |
| • Promotion of healthy diet and physical activity during childhood, adolescence<br>• Maternal/child food supplementation<br>• Maternal/child micronutrient supplementation, including home fortification<br>• Delayed cord clamping<br>• Support for early immediate breastfeeding initiation<br>• Promotion and support for exclusive and continued breastfeeding<br>• Promotion of age-appropriate complementary feeding practices<br>• Management of MAM<br>• Treatment of SAM<br>• Anemia treatment | • Disease prevention & management strategies esp. diarrhea<br>• Family planning & reproductive health services<br>• Maternal mental health support | • Iodized or other MN fortified salt<br>• Staple food fortification<br>• Biofortification and agronomic fortification<br>• Policies to reduce prices or increase access to nutritious foods and diverse diets<br>• Policies to limit marketing of unhealthy foods and BMS including labeling<br>• Promotion of healthy diets and age-appropriate complementary feeding in social protection programs<br>• Nutrition interventions in schools<br>• Nutrition in emergency programs<br>• Mass and social media on nutrition | • Household food security<br>• Poverty alleviation strategies<br>• Women's empowerment<br>• Child protection & support services<br>• Universal education with a gender focus<br>• Early child stimulation<br>• WASH interventions<br>• Food safety<br>• Sugar-sweetened beverage / sin taxes |

**Cross-Cutting Strategies:** Health system strengthening; data system strengthening; community mobilization; monitoring and evaluation for accountability, delivery and implementation approaches for scale, financing

**Figure 15.7: Conceptual Framework for Interventions Related to Child and Maternal Undernutrition. BMS: Breastmilk substitues; MAM: Moderate acure malnutrition; MN: Micronutrient; SAM: Severe acute malnutrition; WASH: Water, sanitation and hygiene.**

interventions within the health sector be considered separately alongside other sectoral approaches, the latter working to improve nutrition in a more supportive manner (see Figure 15.7).

The direct nutrition interventions, which in this framework extend beyond stunting reduction to encompass multiple nutrition outcomes and conditions such as reduction of wasting and anemia, are nearly synonymous with those that were previously named nutrition-specific. However, several approaches that were previously deemed nutrition-sensitive are better reclassified as indirect nutrition strategies (*e.g.*, disease prevention and management, reproductive health). Other interventions, which also contain direct and indirect actions, fall into the other sectoral strategy category [*e.g.*, agriculture and food security, social safety nets (*e.g.*, conditional cash transfers) and other poverty alleviation strategies, promotion of women's empowerment, child protection, education, WASH]. Within this schema, direct interventions address the immediate determinants of all forms of child undernutrition whereas indirect interventions (that are both inside and outside the health sector) work to alleviate the more distal, underlying determinants. Not all indirect interventions explicitly include a nutrition focus, but sectoral agendas (*e.g.*, related to poverty reduction and improvements in education and sanitation) can firmly and positively address underlying determinants of child growth firmly and positively when implemented with a focus on closing social and geographic equity gaps.

## A Roadmap for Reducing Child Stunting at Scale

So, what can a country or region do to reduce child stunting at scale? Based on our case study findings and existing evidence, Figure 15.8 depicts our proposed roadmap to achieve this end, split into 2 phases. In Phase I, we suggest policy and investment cases for reducing child stunting to begin with a robust diagnostic comprised of a situational analysis and stakeholder consultations. The situational analysis should use both quantitative and qualitative data to create a country-level baseline report of levels, trends, and determinants of child stunting.

It would also include an assessment of existing programs and approaches (including new service delivery mechanisms such as results-based approaches, digitally enhanced approaches, and new financing mechanisms such as results-based financing, performance for results, *etc.*), and a range of accompanying policies, legal acts, and actions in other supporting sectors.

A focus on identifying interventions and delivery mechanisms that can ensure equitable reach while addressing capacity and constraints, would be key. For instance, a gap-analysis of what is needed to achieve programmatic scale, including an assessment of financing needs using decision science tools such as Optima Nutrition [26], would be useful.

Findings from the report should be deliberated amongst key stakeholders (*e.g.*, donors, government, nongovernmental organizations, private sector players, *etc.*) to identify gaps and priorities for different intervention packages and delivery strategies in the health/nutrition sector. Phase II involves strengthening of delivery systems and implementation of scaled-up actions identified through the iterative

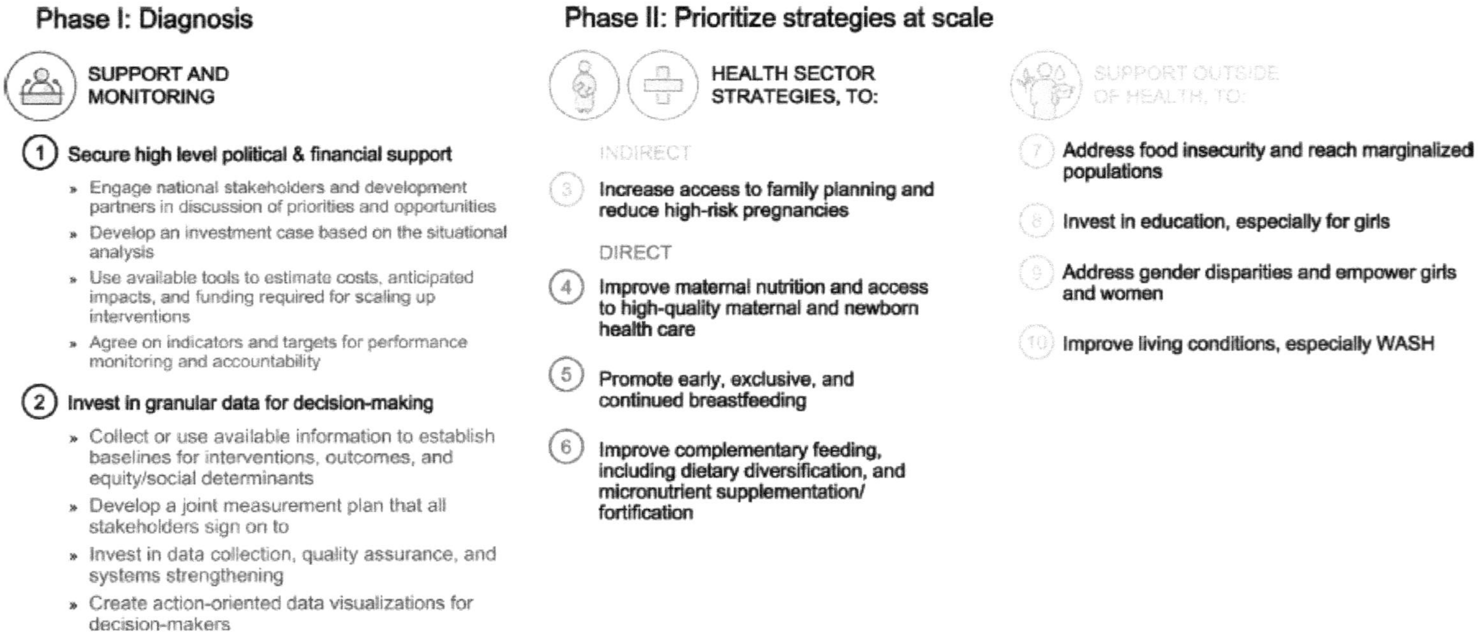

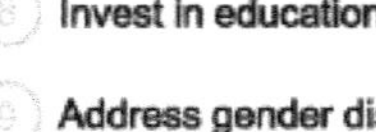
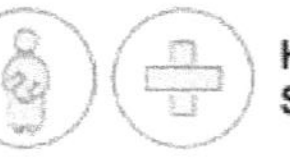

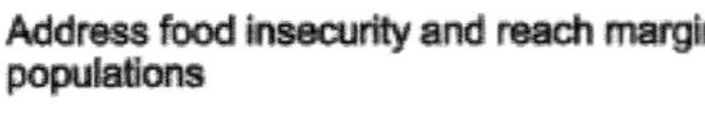

**Figure 15.8: Roadmap to Child Stunting Reducation at Scale. NGO: Non-governmental organization; WASH: Water, sanitation and hygiene.**

process in Phase I in both the health and nutrition sectors and nonhealth sectors. Although we have rank ordered interventions in our framework, each country's adoption of these should depend on its baseline situation, existing status of relevant policies and programs, logistics, capacities, and finances. In addition, policies that shape the broader household and community environment for improvements in child growth require attention such as the extent and nature of economic and agricultural growth, education, and sanitation, with a front and center focus on achieving social equity in these areas. Finally community engagement and buy-in seems critical in many ways. Together, actions across these areas have the potential to accelerate improvements in child growth.

## Summary

This summary article, focused on drivers and initiatives for reducing child stunting at scale, has analyzed global evidence of stunting determinants from LMICs, in-depth case studies from 5 exemplar countries, and the existing nutrition literature base. We found that improvements in child growth faltering and reductions in stunting were related to investments from both within and outside the health sector; we note that ~40 per cent of the impact came from direct and indirect health-related nutrition strategies, 50 per cent from other sectoral strategies, and 10 per cent was unexplained. Notably, improvements in maternal nutrition, maternal education, broad maternal and child health care, and fertility practices yielded some of the greatest gains. However, given differences in context and baseline status of some countries, the relative contribution of various components and the percentage explained by decomposition varied somewhat. In addition, the impact of food security and improved dietary practices (including complementary feeding practices) could be understated in our findings due to limitations in data availability and quality. Lastly, accurate serial coverage data for most direct health sector nutrition interventions (*e.g.*, supplementation) were generally lacking. Taken together, all such observational analyses are subject to ecological fallacies, and hence the relative contribution of various factors to improvements in stunting must be viewed as illustrative and associational rather than deterministic. Additionally, this study did not discuss findings from control or counterfactual countries (*i.e.*, those that have not made significant national stunting reduction) though a second phase of the exemplars project is currently examining large population countries (India, Pakistan, and Nigeria) with the aim of assessing these subnational trends and policy environments more robustly. The quality of anthropometry measures continues to be a challenges in survey datasets; we have examined key metrics of height, weight, and age quality in exemplar country survey datasets and found no major issues in missingness, response mismatch, or digit preference in the included surveys.

The improved classification of nutrition actions presented in this article will allow countries to effectively plan, prioritize, and invest in strategies that will lead to the intended population level nutrition-related outcomes. Given the complex and multicausal nature of suboptimal growth and development, we generally strive for multisectoral approaches. However, clearly demarcating responsibilities by sectors to deliver on their core sectoral mandates with an equity lens also has strong potential to build an enabling environment for nutrition at the household level. Building better action and accountability in terms of resources, finances, leadership, and other factors to enable this can accelerate gains in stunting. It will be critical not to lose the momentum that has been behind the recent global focus on nutrition; for this, we emphasize the importance of and need for strong governance, supportive sectoral strategies, and direct and indirect health interventions for improving nutrition outcomes.

## Future Analyses

These types of holistic analyses can useful to gain a better understanding of country context and nutrition evolution, and will inform future programmatic choices. We suggest that our alternate framework of organizing nutrition interventions as direct/indirect and inside/outside the health sector (in place of nutrition-specific and -sensitive) should be considered when mapping child stunting causal pathways and planning interventions and strategies to accelerate stunting reduction to achieve the 2030 SDGs. Our road map of diagnostics coupled with intervention prioritization could provide a valuable foundation for stunting reduction investment cases at national or regional scales. In addition, we call for a better coordination of the various efforts to understand drivers of nutrition change in a country, and accelerated efforts to undertake these analyses where they have not been done.

## Improving Data

Many aspects of the analysis are dependent on the proxy data that we have on hand as opposed to direct measures of constructs. To illustrate, the interface of maternal and newborn nutrition and its relation to epigenetic factors and fetal growth is well recognized [27]. However, data on important indicators such as maternal prepregnancy BMI, weight gain during pregnancy, and, importantly, gestational age and birth weight are not available. Our exercises ended up deploying a range of proxy indicators in an attempt to better understand the link between maternal nutrition, intrauterine growth, birth weight, and infant growth faltering. With the global drive to improve data capture around birth weight [28], this might improve over time. The same can be stated around disease management, maternal and child diet quality, and coverage of nutrition interventions, where existing survey data are presently insufficient to adequately assess their relative contributions to change over time. Furthermore, existing information on WASH coverage and behaviors does not include important information on frequency, quality, and fidelity of interventions and use, which could be rate-limiting factors in recent trials that have failed to show benefits of hand washing and sanitation interventions alone on linear growth in children [29]. Taking these learnings around data collection and applying them to future studies and programs will be critical. In addition, recent consensus on coverage indicators for evidence based nutrition interventions provides an important opportunity to improve data for future analyses and programmatic and policy direction [30]. Our exercise was also somewhat limited in that we did not have sufficient longitudinal information on population micronutrient status as well as concurrent stunting and wasting [17, 31]. Future studies should be able to identify exemplar countries as those that have seen reductions across all the dimensions of childhood undernutrition. Finally, we did not take into account the concomitant time trends of childhood and adolescent overweight and obesity in the exemplar countries. True exemplars could well be the subset of countries that have seen improvements in undernutrition without the unintended consequences of overweight and obesity at population level, a phenomenon now being observed in most LMICs [32]. In addition to known associations between overweight/obesity and the food environment [33], a few evaluations have shown that food distribution and school feeding programs themselves, intended to address undernutrition, could contribute to the growth in child overweight and obesity [34,35]. Future studies should include these complexities in the mix of determining true exemplars, especially in regions of the world with a growing double burden of malnutrition.

## Policy and Programmatic Way Forward

As highlighted throughout this supplement, the specific drivers of reductions in national-level stunting prevalence for individual countries vary. However, there exist a set of key determinants that must be addressed with policy change and programmatic action to drive further progress on reducing stunting, optimizing child growth and development, and, ultimately, achieving 2030 SDG health and nutrition targets. The opportunities and ambitions to scale up actions in high-burden countries must be matched with financing needs, as assessed by the World Bank, and optimizing available resources (Optima Nutrition) so as to manage expectations of impact. Other collections of country case studies on the key drivers of improved nutrition, such as the "Stories of change in nutrition" [20] and "Stop stunting in South Asia" [36], have highlighted the particular importance of strong leadership, political and institutional commitment to change, and strategic investments targeting the most vulnerable and addressing the biggest risk factors for child growth. Our findings support and concur on the critical importance of these enablers to sustain improvements in child undernutrition. We also stress the vital importance of multisectoral action plans for targeting undernutrition, though the varying importance of sectors will differ based on context. The Scaling Up Nutrition (SUN) movement provides a collaborative and multisectoral framework for governments to commit to ending malnutrition in all its forms [37], but it is important that these plans align with the evidence base. The WHO has also compiled a list of evidence based Essential Nutrition Actions [38] that countries can use to address the immediate determinants of malnutrition, taking into account the local context and progress toward achieving global nutrition targets. At the tail end of the decade of nutrition [39], acceleration of activities within nutrition programs is urgently needed for countries to reach the targets set by the WHA and the SDGs. Our suggested sequence of actions offers a pragmatic evidence-based approach to doing so. Notwithstanding the need for more information and research, we now know enough to move to action at scale, the time for which is now.

## Acknowledgment

Special thanks to Kevin Ho, Dr Oliver Rothschild, and Dr Niranjan Bose from Gates Ventures for funding and overall support to the project. Acknowledgment also goes to Hana Tasic for research support throughout the manuscript preparation.

A heartfelt acknowledgment goes to our research partners and all participants of our research including the national stakeholders and community members that we interviewed. The authors' responsibilities were as follows—ZAB: conceptualized the study and coordinated the research; NA, ECK: initial drafting of the manuscript; ZAB: primary responsibility for the final content; and all authors: contributed to the overall exemplars study design as well as the interpretation of the data, reviewed the paper, and read and approved the final manuscript.

The authors reported no conflicts of interest.

# REFERENCES

1. Victora CG, De Onis M, Hallal PC, Blössner M, Shrimpton R. Worldwide timing of growth faltering: revisiting implications for interventions. Pediatrics 2010;125(3): e473–80.

2. MAL-ED Network Investigators. Childhood stunting in relation to the pre- and postnatal environment during the first 2 years of life: theMALED longitudinal birth cohort study. PLoS Med 2017;14: e1002408.

3. Leroy JL, Ruel M, Habicht J-P, Frongillo EA. Linear growth deficit continues to accumulate beyond the first 1000 days in low- and middleincome countries: global evidence from 51 national surveys. J Nutr 2014;144: 1460–6.

4. WorldHealth Organization.WHOglobal targets 2025: poster [Internet]. 2012 [cited January 16, 2020]. Available from: https: //www.who.int/nutrition/ topics/nutrition_ globaltargets2025/en/.

5. Osgood-Zimmerman A, Millear AI, Stubbs RW, Shields C, Pickering BV, Earl L, Graetz N, Kinyoki DK, Ray SE, Bhatt S, *et al.*, Mapping child growth failure in Africa between 2000 and 2015. Nature 2018;555: 41–7.

6. Black RE, Allen LH, Bhutta ZA, Caulfield LE, de Onis M, Ezzati M, Mathers C, Rivera J. Maternal and child undernutrition: global and regional exposures and health consequences. Lancet 2008;371: 243–60.

7. Black RE, Victora CG,Walker SP, Bhutta ZA, Christian P, De Onis M, Ezzati M, Grantham- Mcgregor S, Katz J, Martorell R, *et al.*, Maternal and child undernutrition and overweight in low-income and middleincome countries. Lancet 2013;382: 427–51.

8. Black MM, Walker SP, Fernald LCH, Andersen CT, DiGirolamo AM, Lu C, McCoy DC, Fink G, Shawar YR, Shiffman J, *et al.*, Early childhood development coming of age: science through the life course. Lancet 2017;389: 77–90.

9. Bhutta ZA, Ahmed T, Black RE, Cousens S, Dewey K, Giugliani E, Haider BA, Kirkwood B, Morris SS, Sachdev HPS, *et al.*, What works? Interventions for maternal and child undernutrition and survival. Lancet 2008;371: 417–40.

10. Bhutta ZA, Das JK, Rizvi A, Gaffey MF,Walker N, Horton S,Webb P, Lartey A, Black RE. Evidence-based interventions for improvement of maternal and child nutrition: what can be done and at what cost? Lancet 2013;382: 452–77.

11. Ruel MT, Alderman H. Nutrition-sensitive interventions and programmes: how can they help to accelerate progress in improving maternal and child nutrition? Lancet 2013;382: 536–51.

12. Das JK, Salam RA, Hadi YB, Sadiq Sheikh S, Bhutta AZ, Weise Prinzo Z, Bhutta ZA. Preventive lipid-based nutrient supplements given with complementary foods to infants and young children 6 to 23 months of age for health, nutrition, and developmental outcomes. Cochrane Database Syst Rev [Internet] 2019;5: CD012611. Available from: https: //www.cochranelibrary.com/cdsr/doi/10.1002/14651858.CD012611.pub3/abstract

13. Smith LC, Haddad L. Reducing child undernutrition: past drivers and priorities for the post-MDG era. World Dev 2015;68: 180–204.

14. Headey D. Developmental drivers of nutritional change: a cross-country analysis. World Dev 2013;42: 76–88.

15. Hossain M, Choudhury N, Adib K, Abdullah B, Mondal P, Jackson AA, Walson J, Ahmed T. Evidence-based approaches to childhood stunting in low and middle income countries: a systematic review. Arch Dis Child 2017;102: 903–9.

16. Leroy JL, Frongillo EA. Perspective: what does stunting really mean? A critical review of the evidence. Adv Nutr 2019;10: 196–204.

17. Khara T, Mwangome M, Ngari M, Dolan C. Children concurrently wasted and stunted: a meta-analysis of prevalence data of children 6–59 months from 84 countries. Matern Child Nutr [Internet] 2018;14: e12516.

18. Martorell R, Young MF. Patterns of stunting and wasting: potential explanatory factors. Adv Nutr 2012;227–33.

19. Vaivada T, Akseer N, Akseer S, Somaskandan A, Stefopulos M, Bhutta ZA. Stunting in Childhood: an overview of global burden, trends, determinants, and drivers of decline. Am J Clin Nutr 2020. In Press

20. Gillespie S, van den BoldM. Stories of change in nutrition: an overview. Glob Food Sec 2017;13: 1–11.

21. Akseer N, Vaivada T, Rothschild O, Ho K, Bhutta ZA. Understanding multifactorial drivers of child stunting reduction in exemplar countries: a mixed-methods approach. Am J Clin Nutr 2020. In Press

22. Gillespie S, van den Bold M. Stories of change in nutrition: a tool pool [Internet]. IFPRI Discussion Paper 01494. Washington (DC): IFPRI; 2015. Available from: http: //ebrary.ifpri.org/cdm/ref/collection/p15738coll2/id/130077.

23. UNICEF. The state of the world's children 1998 [Internet]. New York; 1998. Accessed January 8, 2020. Available from: https: //www.unicef.o rg/sowc98/sowc98.pdf.

24. Shekar M, Kakietek J, Eberwein JD, Walters D. An investment framework for nutrition: reaching the global targets for stunting, anemia, breastfeeding, and wasting. Washington (DC): World Bank Group; 2017.

25. Morris SS, Cogill B. Effective international action against undernutrition: why has it proven so difficult and what can be done to accelerate progress? Lancet 2008;371: 308–21.

26. Optima Nutrition [Internet]. Optima Consortium for Decision Science; 2018 [cited January 16, 2020]. Available from: http: //optimamodel.co m/nutrition/.

27. Katz J, Lee AC, Kozuki N, Lawn JE, Cousens S, Blencowe H, Ezzati M, Bhutta ZA, Marchant T, Willey BA, *et al.*, Mortality risk in preterm and small-for-gestational-age infants in low-income and middle-income countries: a pooled country analysis. Lancet 2013;382: 417–25.

28. Blencowe H, Krasevec J, de Onis M, Black RE, An X, Stevens GA, Borghi E, Hayashi C, Estevez D, Cegolon L, *et al.*, National, regional, and worldwide estimates of low birthweight in 2015, with trends from 2000: a systematic analysis. Lancet Glob Heal [Internet] 2019;7: e 849–60.

29. Pickering AJ, Null C,Winch PJ, Mangwadu G, Arnold BF, Prendergast AJ, Njenga SM, Rahman M, Ntozini R, Benjamin-Chung J, *et al.*, The WASH Benefits and SHINE trials: interpretation ofWASH intervention effects on linear growth and diarrhoea. Lancet Glob Heal [Internet] 2019;7: e1139–46.

30. Gillespie S,Menon P, Heidkamp R, Piwoz E, Rawat R, Munos M, Black R, Hayashi C, Kumar Saha K, Requejo J. Measuring the coverage of nutrition interventions along the continuum of care: time to act at scale. BMJ Glob Heal [Internet] 2019;4: e001290.

31. Wells JCK, Briend A, Boyd EM, Berkely JA, Hall A, Isanaka S, Webb P, Khara T, Dolan C. Beyond wasted and stunted—a major shift to fight child undernutrition. Lancet Child Adolesc Heal 2019;3: 831–4.

32. Popkin BM, Corvalan C, Grummer-Strawn LM. Dynamics of the double burden of malnutrition and the changing nutrition reality. Lancet 2020;395: 65–74.

33. Needham C, Orellana L, Allender S, Sacks G, Blake MR, Strugnell C. Food retail environments in Greater Melbourne 2008–2016: Longitudinal analysis of intra-city variation in density and healthiness of food outlets. J Environ Res Pub Health 2020;17(4): 1321. doi: 10.3390/ijerph17041321.

34. Abarca-Gómez L, Abdeen ZA, Hamid ZA, Abu-Rmeileh NM, Acosta- Cazares B, Acuin C, Adams RJ, Aekplakorn W, Afsana K, Aguilar- Salinas CA, *et al.*, Worldwide trends in body-mass index, underweight, overweight, and obesity from 1975 to 2016: a pooled analysis of 2416 population-based measurement studies in 128.9 million children, adolescents, and adults. Lancet 2017;390: 2627–42.

35. Caballero B, Vorkoper S, Anand N, Rivera JA. Preventing childhood obesity in Latin America: an agenda for regional research and strategic partnerships. Obes Rev 2017;18(Suppl 2): 3–6.

36. Aguayo VM, Menon P. Stop stunting: improving child feeding, women's nutrition and household sanitation in South Asia. Matern Child Nutr [Internet] 2016;12: 3–11. Available from: https: //doi.org/10.1111/mcn.12283.

37. Scaling Up Nutrition. The SUN movement strategy and roadmap [Internet]. 2019 [cited November 18, 2019]. Available from: https: //sc alingupnutrition. org/about-sun/the-sun-movement-strategy/

38. World Health Organizatin. Essential nutrition actions: mainstreaming nutrition through the life-course. Geneva, Switzerland: WHO; 2019.

39. Bhutta Z. Nutrition: howwill the next "Decade ofNutrition" be different from the past one? Nat Rev Gastroenterol Hepatol 2016;13: 441–2.

# Chapter 16

# Role of Livestock in Alleviating the Burden of Hunger by 2030

**B.S. Prakash**

*Sr. Technical Consultant (NRAA) and Former ADG (AN&P),
ICAR, New Delhi*

Eradication of hunger from the world - what is being referred to as zero hunger- by 2030 - is one of the ambitious sustainable development goals (SDGs) of the United Nations. The importance of this SDG - listed at number 2 with eradication of poverty at number 1 position - can be gauged by the fact that its attainment is very crucial for the success of the remaining ambitious 16 SDGs. The core activity for achieving this goal is agriculture production which undoubtedly is intricately linked with livestock productivity. This will necessitate the overall holistic improvement through productivity enhancement of agriculture and livestock with access to technology and markets with international cooperation playing an important role to ensure investments in infrastructure in this sector. According to UNDP, over the past two decades, India's contribution to global economic growth has doubled to almost 15 per cent. Further, income poverty levels have declined, resulting in 133 million people being lifted out of poverty in the past 20 years. However, we have a long way to go with nearly 300 million people still living in extreme poverty. While India produces enough food to feed its population. In 2019 the country was home to nearly 14 per cent of its population which is hungry. A holistic approach to food security requires ensuring available, accessible and nutritious food to eradicate hunger and malnutrition in India. The recent Covid pandemic has added to the woes by large scale migration of several million migrants due to the lockdown imposed in cities and towns across the country seriously affecting economic growth. However, the pandemic has also brought in new opportunities for increasing agriculture productivity - which also includes the allied sector of animal husbandry - with the attention now being given to rural improvement and holds tremendous promise in the years to follow.

Livestock sector, an integral component of India's agricultural economy, has been growing at much faster rate (over 4 per cent) compared to other components of the agriculture sector. It has emerged as an important source of income to the farmers. Moreover, this sector contributes 29 per cent to the national agricultural gross domestic product (GDP) and the demand for animal food products is growing at a rapid pace. India has substantial livestock populations, of goats (149 million), sheep (74 million), cattle (193 million) and buffaloes (110 million). The population numbers of buffaloes are the highest in the world while the other animals - goats, sheep and cattle - standing at the second place. In 2019, India had 535.78 million animal heads. The farmers in India maintain mixed farming system *i.e.* a combination of crop and livestock where the output of one enterprise becomes the input of another enterprise thereby ensuring efficient utilization of resources. Consequently, marginal, small and semi-medium farmers with average operational holdings of area less than 4 Ha own about 85 per cent of the livestock of this country. Since livestock generates income even on a daily basis it provides a guaranteed source of sustenance for the livestock holders. Livestock provides employment in addition to providing nutritious food draught power and manure and also serves as a cushion against crop failure.

## Livestock Sector vs Crop Sector

Rural economy is dependent on livestock because of the inherent benefits such as income, food, manure, draft power as well as social status. Livestock production provides a constant flow of income and reduces the vulnerability of agricultural production. In 2017-18 livestock generated output worth Rs.1043656 crores at current prices in which milk alone accounted for Rs.701530crores (67 per cent). The development path of the Indian economy is manifested various signs of change. The contribution of agriculture sector to total GVA has seen a declining trend, while the contribution of livestock sector to the total GVA and agricultural GVA has increased (Table 16.1).

**Table 16.1: At Current Prices in Rs Crores**

| Year | GVA (Total) | GVA (Agriculture and Allied Sector) | | GVA (Livestock Sector) | |
|---|---|---|---|---|---|
| | | Rs. Crores | Per cent Share to total GVA | Rs. Crores | Per cent Share to total GVA |
| 2011-12 | 81,06,946 | 15,01,947 | 18.50 | 3,27,334 | 4.00 |
| 2012-13 | 92,02,692 | 16,75,107 | 18.20 | 3,68,823 | 4.00 |
| 2013-14 | 103,63,153 | 19,26,372 | 18.60 | 4,22,733 | 4.10 |
| 2014-15 | 115,04,279 | 20,93,612 | 18.20 | 5,10,411 | 4.40 |
| 2015-16 | 125,66,646 | 22,25,368 | 17.70 | 5,84,070 | 4.60 |
| 2016-17 | 138,41,591 | 24,84,005 | 17.90 | 6,39,912 | 4.60 |

The contribution of livestock to the total output of the agriculture sector increased from 19 per cent in 1983 to 28.4 per cent in 2019. Driving livestock growth, are changes in the utility of livestock for farmers and in food consumption pattern. Consumption of livestock products like eggs, milk and meat is increasing due to rise in the income of the booming middle class, both in urban and rural areas. The value of output from milk group has been higher than the same for three major crops *viz.*, paddy, wheat and sugarcane making milk the largest agricultural commodity in India.

**Table 16.2: Trends in Real Value of Output (2011-12 prices) Rs. in Crores**

| Item | 2011-12 | 2012-13 | 2013-14 | 2014-15 | 2015-16 |
|---|---|---|---|---|---|
| Paddy | 170595 | 169773 | 172144 | 170369 | 168028 |
| Wheat | 118068 | 115884 | 119051 | 107828 | 114880 |
| Sugarcane | 60139 | 58614 | 56181 | 59464 | 54179 |
| Milk group | 327767 | 339240 | 352247 | 374269 | 396691 |

Vegetable proteins, mainly in cereals, legumes and vegetables are of poorer quality than animal proteins, not only because of their lower digestibility but also because they are limiting in one or more of the essential amino acids of human requirement. Cereal proteins are generally deficient in lysine and pulses or legume proteins contain low amounts of methionine. However, when both cereal and pulses (legumes) are present in the diet in proper proportions, the proteins from these two sources supplement each other and make good each other's deficiencies in lysine or methionine to a significant extent. In spite of such supplementation, however, Indian diets deriving their proteins from cereals and pulse combination (5:1) have a biological value of only 65 per cent relative to egg protein.

## Role of Dairy Sector in National Economy

India ranks first among the world's milk producing nations, with a sustained growth in the availability of milk and milk products for growing population. The annual milk production was 187.7 million tonnes in 2018-19 recording an annual growth rate exceeding 6 per cent per cent with a per capita milk consumption of 394 g. The Indian dairy sector is characterized more by 'production by masses' than 'mass production'. Unlike leading milk producing countries in the world, large proportion (95 per cent) of milk producers in the country hold only 1 to 5 animals per household.

Currently, about 54 per cent of the milk produced in the country is surplus for domestic marketing out of which about 38 per cent is handled by the organised sector (*i.e.* about 20 per cent of total milk production), equally shared by Co-operatives and Private dairy organizations. Almost 46 per cent of milk produced in India is retained by the households or sold to nonproducers in the rural area. Unlike in most developed nations where 90 per cent of the surplus milk is processed through organised sector, it is the unorganised sector that predominates the marketing and processing of milk segment in India. In the interest of economic well being of the milk producers, it is essential to provide rural milk producers with greater access to the organised sector. This would not only ensure remunerative prices to farmers for their produce but also encourage more farmers to adopt dairying as a source of livelihood. Co-operative sector plays a major role in supplying liquid milk in the domestic market. Until March 2018, a total of 166 lakh members were covered under Dairy Co-operatives covering about 1.9 lac villages and 476 lakh kg milk per day collection. The huge surplus milk being marketed under the unorganized sector needs to be brought under the gambit of cooperatives by bringing another 2 lac villages over the next decade. The dairy sector is poised for a tremendous growth and a target of 270 million tonnes of milk with a per capita consumption of 500 g per day should be achievable by 2030.

## Meat Sector

Meat is a valuable commodity and an important source of protein. The meat availability in India is only about 15g/person/day against the ICMR (Indian Council of Medical Research) recommendation of 30g/person/day (Islam *et al.*, 2016). Production of meat in India increased from 1.9 million MT in 1988-89 to 8.11 million MT in 2018-19. In 2009-10, next to poultry, goat accounted for the greatest share (17 per cent) of total meat production, while buffalo's share stood at 14 per cent. However, the share of buffalo meat increased to 19 per cent in 2018-19, while that of goat decreased to about 14 per cent. This provides some indications in respect of the increasing demand for buffalo meat in the country. Of the agri-exports from India, buffalo meat is one of the major items. India's annual foreign exchange earnings from buffalo meat export went up from Rs. 3,500 crore in 2007-08 to Rs. 27, 000 crore in 2015-16. It is estimated that this sector has the potential to reach a target of Rs. 40, 000 crore as foreign exchange earnings in the next five years. Buffalo meat export has experienced a phenomenal growth since the year 2000, when India accounted for about 5 per cent of the total global trade. By 2015, India emerged as the largest exporter in the world comprising 21 per cent of the world's carabeef export. With the present per capita meat consumption annually of 5.5 kg it is envisaged to increase to 12.1 kg by 2030 with growth promoting interventions (ICAR-NRC Meat Vision document).

## Goat and Sheep Rearing

Goats and sheep are raised mainly for meat, milk, and skin and provide a flexible financial reserve in bad crop years for the rural population. About 5 million families in India are engaged in various activities relating to rearing of small ruminants. Farmers of rainfed dry regions prefer rearing of small ruminants because they are less expensive to purchase and require minimal inputs and maintenance costs besides being climatic stress resilient. An added advantage is their higher prolificacy which makes it easy for their disposal.

Rearing of small ruminants is more profitable with assured and continuous income. Small ruminants, require less forage than large cattle, and an acre of pasture land can sustain five times as many goats than cattle. These animals forage on plant varieties that are not preferred by other grazing animals can improve the quality of pasture land. They are often more manageable, affordable and hardy for the small holder farmer. Goats produce milk and meat, and diverse by-products (cheese, skin, cosmetics, gut string, *etc.*). Indian sheep breeds are not considered as pure milch breeds. Sheep, however, additionally also produce wool. There is a growing demand and opportunity for multi-dimensional growth of the sector to meet the robust demand for high value goat meat and milk products in domestic and global market. India is the largest exporter of sheep and goat meat to the world. The country has exported 21,950.71 MT of sheep and goat meat to the world for the worth of Rs.837.76 crores during the year 2015-16 though the exports dropped marginally to about Rs. 800 crores in 2018-19.

## Piggery

Piggery is a popular activity, with no social taboo in the north eastern states including Sikkim. However, this sector has not been able to flourish as an enterprise, since it is mainly the nondescript breeds having low productivity rate, that are reared. The efforts to introduce exotic germplasm have not met success as yet, and indiscriminate breeding among animals of inferior quality has continued. Inbreeding has become common. Further, quality animals have ended up in slaughter houses. The ICAR-National Research Centre on Pigs, Guwahati, is developing hybrid high yielding pigs with local crosses with exotic germplasm and disseminating it among the livestock holders. Total pig population in the country declined by 12 per cent from 2012 count to 9.06 million in 2019 (Livestock census 2019).

## Poultry

India possesses an estimated 850 million poultry birds. Chicken varieties like Vanaraja, Gramapriya and Srinidhi *etc.* suitable for backyard free range farming have been developed by ICAR - Directorate of Poultry Research, Hyderabad which are well accepted by the rural and tribal people cutting across the different sectors, religions, *etc.*, all over the country. These chicken varieties with multi-coloured plumage resemble the native chicken in their feather pattern and produce more meat and eggs than the natives. The improved varieties of birds have gained wider acceptability across the country. These are generally maintained by the women with household waste, green fodder, insects, supplementary feed, *etc.* allowing the women to live respectful life with additional earning to the family.

Poultry market is currently dominated by meat of broiler birds, whereas earlier chicken implied "spent hens" or male birds. Broiler birds made in roads into India around 1975, when hatcheries imported the parents of the hybrid broilers. Poultry sector is the highest with a 50 per cent meat contribution with buffalo meat coming second at 19 per cent.

Annual production of eggs has grown from 183 crore eggs in 1950-51 to 8,814 crore eggs in 2016-17 and 10330 crore eggs in 2018-19 with a per capita consumption of 79 eggs. Hence, it is necessary to increase the production of eggs manifold to ensure an average daily egg consumption. Nationally, around 19 per cent of the egg production is from backyard poultry enterprises, in which 64 per cent are produced by indigenous (desi) fowl. Ducks contribute about 7 per cent of the eggs in this sector. The organised or commercial sector contributes about 81 per cent of the eggs produced. The large commercial poultry layer farms are mainly concentrated in South India with Andhra Pradesh, Tamil Nadu and Telangana representing nearly 60 per cent of the egg production. Commercial ventures also source eggs from small holders, regularly or on contract, and are a channel for the small farmers to access the larger market. However, in backyard poultry West Bengal is a major contributor with nearly 30 per cent of the total egg production followed by Kerala providing around 14 per cent. While backyard poultry farms are prevalent across the country, they are highly dependent on market accessibility.

## Challenges in Livestock Sector

Although the livestock sector in India is on a growth trajectory the following challenges need to be addressed for ensuring a faster growth in this sector not only to meet the growing urban demand for livestock products but also to make them affordable to potentially meet the zero hunger target by 2030.

### Low per Capita Milk Production

The key factor for high milk production of cows and buffaloes is the genetic merit of the animal for milk production. The indigenous cattle genetic resources of India are represented by only 43 recognized breeds that account for 20 per cent of the cattle population. Another 21 per cent cattle population is crossbred, that means 59 per cent of the cattle in Indian are non-descript. The stock of female breedable non-descript cattle is over 40 million, whose average age at first calving is about 60 months, with a milk production average of just about 500 kg in a lactation and a long calving interval of 20 to 24 months. In case of buffaloes also, about 43 per cent are non-descript animals with very low genetic potential. The critical challenge for breed improvement programs must target identification and improvement of these non-descript animals as well as reduce age at maturity and improve lactation through scientific breeding.

There is need to prioritize cattle and buffalo breeds as per their economic importance. This will be helpful to farmers in order to increase their income through rearing of livestock and sale of milk, milk products and animals as well. The breeding bull contributes significantly in enhancing the genetic potential of its progenies for economically important traits, like milk production, fertility, body conformation *etc*. Genetic improvement for a specific trait or group of traits largely depends on how effectively these traits are recorded and what accuracy of selection is obtained for production of bulls. Currently, about 3800 bulls are used at various semen stations for producing semen with replacement of about 940 bulls each year. AI is the preferred method for attaining genetic progress in milch animals. To accelerate the genetic progress, the proportion of milch animals bred through AI needs to be raised substantially. Presently, only about 30 per cent of the breedable animals are bred through AI and 70 per cent are bred through natural service. It is imperative that by 2030 we should be able to achieve 100 per cent AI for which there needs to be production of quality cattle and buffalo males through large scale application of modern embryo biotechniques including embryo transfer and cloning and improved reproductive technologies for efficient fertility and reduced calving intervals. Promising indications are available from various research Institutions as well as private farms where modern embryo biotechnologies are being used for production of high genetic males.

## Shortage of Feed and Fodder

Feed is the single largest input accounting for 60-70 per cent of the cost of production. Feed shortage has been identified as one of the major factors responsible for low productivity in Indian livestock. Although there is variation among the published reports with regard to the extent of deficit, one consistent finding across all the studies is deficit of the all major class of feed resources namely- dry fodder, green fodder and concentrates. The large Indian land mass has among other things to generate nutritional support to a huge animal (536 million), poultry (850 million), human (1,352 million) populations. The three main sources of forage supply are crop residues, cultivated fodders and forage from common property resources like forests, permanent pastures and grazing lands which are presently contributing 54, 28 and 18 per cent, respectively. The projected livestock demand for dry fodder, green fodder and concentrate for 2020 is 468, 213 and 81 mt on dry matter basis, whereas the availability is estimated to fall short by 11, 35 and 45 per cent, respectively.

Further it is projected that the extent of deficit would widen in the near future due to growing demand for the livestock products and the shrinking resource base of the feed resources. Presently the dry fodder constitutes the major source of the feeds accounting for > 60 per cent of the resources followed by the greens and the concentrates. Availability of dry fodder and concentrates comprising of grains, grain by-products and oil cakes of food crops are directly proportional to their food grains production. Plateau in the productivity of major food crops and with no further scope of increasing the cropped area, the scope of further improving the availability of crop residues and concentrates is very limited. Similarly the area under fodder crops has remained static over the last two decades and the scope of increasing the area under fodder crops is limited due to the growing competition for land and water resources for food/commercial crops. Under the present circumstances improving the productivity of food crops and green fodder is one of the potential options for improving the feed resource availability. Besides these, utilizing the available feed resources judiciously by proper balancing the diets and management of the resources are other promising options. Under these approaches ration balancing, compound concentrate mixtures, strategic supplementation, fodder bailing, densified feed blocks, total mixed rations, fodder conservation-hay/silage making are some of the possible options.

Agroforestry is a viable option which presently contributes 9-11 per cent to fodder requirement of livestock. During dry season, forage from trees may constitute the only source of green forage particularly in arid, semi-arid and hilly regions. The importance of tree fodder increases with the severity of drought and progression of drought season. Tree leaf fodder provides 50-90 per cent of the forage demand during lean periods. Leaf fodder from top feeds is rich in essential nutrients as compared to grass or crop residues. More than 200 tree species have forage value and contain more than 55 per cent total digestible nutrients (TDN) and are, thus, a better energy source. Assuming availability of another 10 per cent of community wastelands for silvipasture, the additional fodder production (@ 2.0 t/ha/yr at 40 per cent of research yield) would be 12 million tons.

## Animal Health Care

The animal health services in the country are provided through a network of over 65000 veterinary hospitals/dispensaries and aid centres (comprising stockmen centres and mobile dispensaries) under the State Animal Husbandry departments. In addition, to a limited extent, some cooperatives, NGOs and private practitioners participate in the provision of these services. Despite the vast network of veterinary institutions, the incidence of diseases and epidemics is very high in the country that impinges on the productivity enhancement. The poor status of animal health stems from the extremely limited attention paid to preventive health care services and inefficiencies in provision of curative health services. Most of the government dispensaries, hospitals and AI centres are stationary and are primarily engaged in providing curative health cover and breeding services. In the case of curative health services, a large gap also exists between the need and the availability of cow-side animal health service providers.

## Increasing Dairy Cooperatives

Currently, Dairy cooperative societies along with Producer Companies cover about 1.9 lac villages. There are about 3.2 lakh villages which have milk production above 200 kg per day at present. However, with the increasing demand of milk and milk products it is likely that the number of villages with milk production potential for dairying will increase to around 4 lacs in the next few years. Thus, about 2 lac villages will have potential for dairying and need to be covered by Dairy Cooperatives or Producer Companies during the next 10 years. Therefore, there is need to expand coverage of cooperatives to new uncovered areas so as to bring more milk producers under the fold of organised sector, thereby providing rural milk producers with greater access to organised milk processing sector.

## Expanding Abattoirs

There is need for developing and modernising municipal abattoirs and retail outlets for slaughter and sale of small ruminant and buffalo meat. The Ministry of Food Processing Industries (MOFPI) has schemes for setting up of new/modernisation of existing abattoirs with a view to ensure scientific and hygienic slaughtering of the animals and supply of quality meat and meat products.

## Small Ruminant Improvement

The majority of goat and sheep in India are non-descript breeds or they are 'local' animals not necessarily belonging to a distinct, 'recognized' group or breed. There is also a lack of good quality data on small ruminant numbers, breed population, trends over time and drivers of these trends. Poor breeding practices due to lack of awareness of scientific breeding practices as well as inter-mating among available animals is common due to limited availability of good quality bucks and rams. The current pattern of producing low genetic merit animals is due to little selection process of males used for breeding.

Vaccines for major diseases are either not available in adequate numbers, or cold chain supply systems are not adequately maintained, affecting the efficacy of the product.

An erosion of synergies between agriculture and animal husbandry due to an interplay of various factors, makes small holder livelihoods more vulnerable. Factors like loss of designated grazing lands, closure of forest areas, and diversion of common lands for other purposes has intensified pressure on available resources, and their degradation. Encroachment of common lands, and increasing industrialization have adversely affected small ruminant rearing, which used to be a highly productive low-input system, dependent on these lands for fodder and grazing resources.

## Government and Non-Government Initiatives for Meeting the Challenges

There have been several government initiatives for promoting livestock development in the country for the improvement of production of cattle, goat, sheep, pig and poultry in the country. Schemes and programs are rolled out by various agencies of Central and State Governments, Dairy Cooperative Federations, NGOs, corporate houses *etc.* to support dairy development and animal husbandry in the country.

### National Livestock Mission (NLM)

The National Livestock Mission (NLM) of the Government commenced from 2014-15 to cover all the activities required to ensure quantitative and qualitative improvement in livestock production systems and capacity building of all stakeholders. This Mission is formulated with the objective of sustainable development of livestock sector, focusing on improving availability of quality feed and fodder. This mission is implemented in all the states. The Mission has four sub-missions; namely, Fodder and Feed Development, Livestock Development, Pig Development in North-Eastern Region and Skill Development, Technology Transfer and Extension. The second sub-mission on Livestock Development has provisions for productivity enhancement, entrepreneurship development and employment generation (bankable projects), strengthening of infrastructure of state farms with respect to modernisation, automation and bio-security, conservation of threatened breeds, minor livestock development, rural slaughter houses, fallen animals and livestock insurance.

### National Programme for Bovine Breeding and Dairy Development (NPBBD)

This program has been initiated in February 2014 by merging four ongoing schemes of the Department of Animal Husbandry and Dairying in the dairy sector, *viz.*, National Project for Cattle and Buffalo Breeding (NPCBB), Intensive Dairy Development Programme (IDDP), Strengthening Infrastructure for Quality and Clean Milk Production (SIQ and CMP) and Assistance to Cooperatives (A-C). This has been done with a view to integrating milk production and dairying activities in a scientific and holistic manner, so as to attain higher levels of milk production and productivity to meet the increasing demand for milk in the country. The scheme has two components (a) National Programme for Bovine Breeding (NPBB) and (b) National Programme for Dairy Development (NPDD). The scheme implementation emphasizes on arranging quality AI services at farmers' doorstep; to bring all breedable females under organised breeding through AI or natural service using germplasm of high genetic merit; to conserve, develop and multiply selected indigenous bovine breeds of high socio-economic importance; to provide quality breeding inputs in breeding tracts of important indigenous breeds so as to prevent the breeds from deterioration and extinction.

## Govt. of India's Rashtriya Gokul Mission (RGM)

This is a focussed project under National Programme for Bovine Breeding and Dairy Development. The Mission also envisages establishment of integrated cattle development centres, Gokul Grams to develop indigenous breeds including upto 40 per cent nondescript breeds. It is implemented through State Implementing Agencies (SIA) *viz.* Livestock Development Boards.

## National Mission on Bovine Productivity (NMBP)

Government of India rolled out National Mission on Bovine Productivity scheme in November 2016 with the objective to enhance milk production and productivity of indigenous bovine breeds and thereby making animal husbandry more remunerative to the farmers. The scheme comprises following components. – *Pashu Sanjivini: Pashu Sanjivini* is an animal wellness programme encompassing provision of animal health cards ('Nakul Swasthya Patra) along with an unique identification number (UID) for animals in milk and uploading data base. Under the scheme, as many as 9 crore number of animals in milk are being identified by assigning UID and further the data is being uploaded in the INAPH database. The health card contains information on vaccination, deworming and treatment details.

Advanced Reproductive Technology: Assisted Reproductive Technique is being used to improve availability of disease free indigenous female bovines through sex sorted technology. The scheme envisages creation of facility for production of 16 million doses annually. Under the scheme, 50 embryo transfer technology labs with *In vitro* Fertilization capacity are being established for exponential multiplication of elite animals of indigenous breeds.

## E Pashuhaat

E Pashuhaat is a portal developed for connecting breeders and farmers of indigenous breeds regarding availability of quality bovine germplasm. The farmers/ breeders may sell or purchase their breeding stock through the portal.

## National Bovine Genomic Centre (NBGC)

National Bovine Genomic Centre has been established with the objective to increase milk production and productivity of indigenous cattle. By using genomic selection indigenous bovine breeds can be made viable within few generations.

# National Dairy Plan (NDP)

This is a Central Sector Scheme being implemented by the National Dairy Development Board (NDDB) through End Implementing Agencies (EIA). NDP is a scientifically planned multi-state initiative. Its various components are:

Productivity Enhancement: This component aims at increasing bovine productivity through a scientific approach to animal breeding and nutrition. The expected results from the interventions proposed under this component are mainly increased milk production through increased productivity per milk animal, increase in number of milk animals, improved conception rates, improved animal nutrition, reduction in feeding costs per kg of milk produced and reduction in methane release per kg of milk produced by animals covered under Ration Balancing Program (RBP). The scheme has the following components:

## 1. Village based Milk Procurement Systems

Efforts to increase milk production through increase in productivity would need to be supported by expanding the network of village based milk procurement systems to collect milk in a fair and transparent manner and ensure timely payments. Investments in village level infrastructure for milk collection and bulking such as milk cans, bulk milk coolers for a cluster of villages, associated weighing and testing equipment and related IT equipment would be made available. The important expected results from the interventions proposed under this initiative are an increase in the number of additional villages covered and more milk producers organised into Dairy Cooperative Societies and Milk Producer Institutions. The organised sector is still small and needs to grow in the interest of farmers.

## 2. Project Management and Learning

The main results expected under this component are effective coordination of project activities among various EIAs, timely preparation and implementation of annual plans, regular review and reporting of project progress and results, a comprehensive and functional project management information system (MIS) and learning that will support improvement and innovation. Importantly, it will also facilitate the development of the skills and knowledge of personnel involved in the implementation of the project and develops capabilities for enhanced capacity building which would extend beyond the life of the project.

# Dairy Entrepreneurship Development Scheme

The Department of Animal Husbandry and Dairying (DAHD), launched a pilot scheme entitled "Venture Capital Scheme for Dairy and Poultry" in the year 2005-06. The main objective of the scheme was to extend assistance for setting up small dairy farms and other components to bring structural changes in the dairy sector. During a mid-term evaluation of the scheme, certain recommendations were made to accelerate the pace of implementation of the scheme. Taking into account the recommendations of the evaluation study and the representations received from various quarters including the farmers, state governments and banks, DAHD decided to make some key changes to the scheme, including changing its name to

Dairy Entrepreneurship Development Scheme (DEDS). The revised scheme has come into operation with effect from September 2010. The objectives of the scheme is to promote setting up of modern dairy farms for production of clean milk, to encourage heifer calf rearing, thereby conserving good breeding stock, to bring structural changes in the unorganised sector so that initial processing of milk can be taken up at the village level itself, to upgrade the quality and traditional technology to handle milk on a commercial scale and to generate self-employment and provide infrastructure mainly for unorganised sector, farmers, individual entrepreneurs, NGOs, companies, groups of organised and unorganised sectors, *etc.* Groups of organised sectors include Self-help Groups (SHGs), dairy cooperative societies, milk unions, milk federations, *etc.* and they can take advantage of this scheme.

## Dairy Processing and Infrastructure Development Fund (DPIDF)

Government of India has approved creation of 'Dairy Processing and Infrastructure Development Fund (DIDF) under Central Sector Scheme in December 2017. The components of the scheme are: a) modernisation and creation of new milk processing facilities, b) manufacturing facilities for value added products, c) milk chilling infrastructure, d) setting up electronic milk testing equipment e) project management and learning, and, f) any other activity related to the dairy sector targeted to contribute to the objectives of DIDF, as decided by Government of India in consultation with the stakeholders.

## Strategic Productivity Enhancement Programme (SPEP)

This scheme was initiated by Gujarat Cooperative Milk Marketing Federation (GCMMF) with an aim to develop genetically improved animals with high productivity. It begins with the selection of elite animals, to develop pure bred cows and buffaloes with high genetic potential. This programme comprises three sub programmes which can be implemented across India with certain objectives like registration of all animals covered under AI, upgradation and improving productivity of state specific cattle and buffalo breeds, elite/superior animal identification, registration and breeding with high genetic merit semen, milk recording of elite/superior animals for bull mother farm, genetic up gradation of non-descript animal breeds and calf rearing programme. GCMMF has already implemented Strategic Calf Rearing Programme with objectives to achieve reduced age at first calving, inter-calving interval and improved productive life of dairy animals.

## Other Initiatives

Initiatives for the genetic improvement of animals include, establishment of Cattle Breeding Farms, National Kamadhenu Breeding Centre (NKBC), Central Herd Registration Scheme (CHRS) and Central Frozen Semen Production and Training Institute (CFSPTI). Seven central cattle breeding farms have been established for production and supply of high genetic merit bulls of indigenous breeds (Tharparkar and Red Sindhi cattle breeds and Murrah and Surti buffalo breeds). National Kamadhenu Breeding Centres (NKBC) are set up as Centres of Excellence. A nucleus herd of all the indigenous bovine breeds will be conserved and developed with the aim of enhancing their productivity and upgrading their genetic makeup. The NKBC will act as repository of germplasm of all indigenous breeds and supply certified germplasm to the farmers undertaking rearing of indigenous breeds. Under the CHRS, four units have been established for identification and propagation of indigenous bovine breeds (Gir, Kankrej, Hariana and Ongole cattle breeds and Murrah, Mehsani, Jaffarabadi and Surti buffaloes). The Central Frozen Semen Production and Training Institute is undertaking production and supply of semen doses of high genetic merit bulls of indigenous breeds (Red Sindhi, Tharparkar and Murrah buffalo).

In the field of animal nutrition, besides Ration Balancing Program other important initiatives are Mineral Mapping Programme and Quality Mark. The NDDB has done mineral mapping in most of the major dairying states to improve both productivity and productive life; and supplementation of area specific mineral mixture in ration of dairy animals. Also NDDB has initiated implementation of common 'Quality Mark' for various variants of cattle feed and mineral mixtures manufactured in the cooperative, government/semi-governments sector.

In order to reduce morbidity and mortality, various efforts are made by governments to provide better health care through polyclinics/veterinary hospitals, dispensaries and First Aid Centres including mobile veterinary dispensaries. Animal Quarantine and Certification Service (AQCS) stations are setup to prevent the ingress of dangerous exotic diseases and diseases of national importance into the country through imported livestock and livestock products. Zoonosis and one health policy of WHO are also the components of AQCS. National Veterinary Biological Products Quality Control Centre has been established at Baghpat, Uttar Pradesh to undertake the quality control and assurance of standard, efficient and safe veterinary biologicals in India and to act as a nodal institute to recommend licensing of veterinary vaccines in the country with a vision to promote healthy and productive livestock in Indian subcontinent using standard, efficient and safe veterinary biologicals. One central and five regional disease diagnostic laboratories have been set up in the country by the government in order to provide referral services over and above the 250 existing disease diagnostic laboratories. The Centre for Animal Disease Research and Diagnosis of ICAR - Indian Veterinary Research Institute, Izatnagar is functioning as central laboratory. National Project on Rinderpest Surveillance and Monitoring system has been set up with the objective of strengthening the veterinary services to maintain required vigil to sustain the country's freedom from Rinderpest and Contagious Bovine Pleuropneumonia infection secured in May, 2006 and May, 2007 respectively. Under the programme, surveillance of various animal disease including syndromic diseases with more focus on Contagious Bovine Pleuropneumonia and Bovine Spongiform Encephalopathy are being undertaken in the country to maintain India's freedom status from these diseases. Foot and Mouth Disease Control Program is being implemented to prevent economic losses due to Foot and Mouth Disease and to develop herd immunity in cloven footed animals. National Animal Disease Referral Expert System (NADRES) is a web based information technology system developed in order to streamline the system of animal disease reporting from the states/UTs. The main objective of NADRES system is to record and monitor livestock disease situation in the country with a view to initiate preventive and curative action in a timely and speedy manner.

## Genetic Improvement in Sheep and Goat (GISC)

In order to enhance the incomes from livestock, one of the critical needs is to achieve higher productivity of meat, milk and wool. This calls for the genetic improvement of indigenous breeds, as well as non-descript small ruminants. Based on a gap analysis of productivity parameters, a two step Action Plan with focus on separately identified breeds for meat, milk and wool production is proposed. These steps are: i. Genetic improvement of identified indigenous descript breeds of sheep and goat through selective breeding for better yielding breed stock for meat, milk and wool. ii. Genetic improvement of non-descript breeds from existing improved descript indigenous breeds. Accordingly, a breeding plan has been formulated to fulfil the twin goals for Genetic Improvement in Sheep and Goat. This is being implemented in a comprehensive manner.

## Wool Marketing Scheme (WMS)

To bring greater focus on marketing of raw wool, the Ministry of Textiles has introduced a new scheme, namely, Wool Marketing Scheme (WMS) in all major wool producing states to support greater procurement of wool on remunerative price by creation of Revolving Fund for Marketing of Wool. Another activity supported are operation of e-portal for auction of wool, formation of wool producers' societies, financial assistance to strengthening infrastructure required for marketing of existing wool mandi/grading centres (storage halls, auction facility, testing platform *etc.*).

## Animal Husbandry Infrastructure Development Fund (AHIDF)

The infrastructure needs of the animal husbandry sector including sheep, goat, poultry and piggery gets a big push with the Union Budget, 2018 announcement of creating a Corpus Fund of the size of Rs. 2,450 crores. The objectives of AHIDF are to address the inadequacy of animal husbandry infrastructure at different levels with financial leveraging for doubling of farmers income and increasing availability and accessibility to protein-rich food to the young population. Proposed activities are establishment/strengthening of livestock breeding farms, poultry breeding farms, poultry hatcheries, establishment/strengthening of semen stations for goat, sheep and pig, establishment/strengthening of Skill Training Institutes (with residential facility), establishment/strengthening of feed mixing units, by-pass protein units, fodder block making unit, veterinary healthcare centre, dispensaries, biological units for healthcare, catering to the need for upgrading the forward linkages like livestock and livestock product markets, retail outlets, cold-chain connectivity, *etc.*

## Innovative Pig Development Projects for North-East (IPDPNE)

In order to check the negative trend of growth rate in pig population and provide an impetus to piggery, a sub-mission, "Innovative Pig Development Project for North East (IPDPNE)" has been rolled out under the 'National Livestock Mission (NLM)'. The challenges of this sector lie in bridging the gap that exists between the current low productivity index of the Indian breeds and the potential growth rate and mature weight. The objective of IPDPNE is to increase the pig meat production and to increase the income of the pig rearing farmer/entrepreneur/ NGO/Cooperative Society, *etc*. This is envisaged to be achieved by importing and incorporating superior germplasm of high genetic merit. The components of this initiative include strengthening of pig breeding farmers, import of germplasm, support breeding programmes, propagate reproductive technologies, and health cover. Appropriate measurable outcomes, including income growth of farmers, are being developed and monitored.

## Innovative Poultry Productivity Project (IPPP)

Private industry and NABARD generally encourage economically large scale poultry which are viable/bankable projects. However, these remain beyond the reach of the small and marginal farmers and resource poor, and they need to be supported. Accordingly special focus is required to upscale small poultry farmers and develop suitable linkages with larger processing industry. Presently, under National Livestock Mission (NLM), there exists a scheme called, 'Rural Backyard Poultry Development (RBPD), which covers Below Poverty Line (BPL) families to enable them to gain supplementary income and nutritional support. Under RBPD, chicks/birds suitable for raising in the backyard are reared in the mother units, for upto four (4) weeks, and are then distributed to the beneficiaries spread over at least two batches. It is possible to move incrementally from this subsistence model of backyard poultry farming to an entrepreneur model comprising 200-400 birds. In case of low-input technology (LIT) birds, it would be possible to later upscale to 1,000-2,000 birds for larger scale poultry farming. A similar model is also feasible in case of broilers, where small scale broilers set up initially can be scaled up to commercial scale. It would help to promote small scale broiler in a cluster mode. This would help in providing various services, including market linkages in an efficient manner.

## Necessary to make the Country Drought Resilient

Rainfed agriculture contributes more than 45 per cent of country's food grain production and supports more than two thirds of livestock population and thus is considered critical to food security, equity and sustainability. More than 50 per cent of net sown area is under rainfed regions which are highly vulnerable to various climate risk factors. The major challenges in rainfed agriculture are-decline in the factor productivity, poor soil health, loss of soil organic carbon (SOC), ground and surface water pollution, water related stress, increased incidence of pests and diseases, increased cost of inputs, decline in farm profits and the adverse impact of climate change. Poor infrastructure, credit and support, lack of insurance, lack of skill and knowledge further aggravate the vulnerability of rainfed systems. Most rainfed areas are confined to arid, semi-arid and sub humid regions of the country and are highly dependent on rainfall for their agriculture and allied activities. Though country has been receiving 4000 BCM of water through rainfall annually, it has been facing severe water crisis over past decades due to erratic or uneven distribution of rainfall. Climate change, mismanagement of agriculture practices, non-judicious use of groundwater water resources, lack of soil moisture conservation practices further aggravate this condition. Hence, drought proofing of rainfed areas from drought is highly important.

National Rainfed Area Authority (NRAA) established in 2006 as an expert body of Ministry of Agriculture to provide the much needed knowledge inputs regarding systematic up-gradation and management of country's dryland and rainfed agriculture aims to enhance prosperity, development and inclusiveness of farmers in rainfed areas on a sustainable basis. NRAA, in association with CRIDA and other DAC and FW and ICAR institutions, is undertaking development of action plans for drought proofing of about 150 vulnerable drought-prone districts in the country. The different interventions for livestock resilience will play a big role for mitigating the losses to the farmers while preparing them for meeting adversities due to drought.

## Ensuring Continuance of the Tradition of Nomadic Pastoralism

There is very little talk in all policies advocated till date on improving pastoralism - which holds tremendous advantages in the national context for increasing livestock productivity.

As mentioned before, the projected demand for dry fodder, green fodder and concentrate for 2020 is 468, 213 and 81 million tonnes on dry matter basis, whereas the availability is estimated to fall short by 11, 35 and 45 per cent, respectively. Herein lies the importance of supporting the ancient tradition of nomadic pastoralism which reduces the pressure on feed and fodder resources for the livestock being reared by them.

Most pastoralists are nomadic or migratory. It is believed that the practice of nomadic or migratory pastoralism started as a result of Neolithic revolution, also known as the first agricultural revolution. During that time, humans had just managed to domesticate some animals and therefore moved with them to places with green pastures. Nomadic pastoralism is a tradition among several communities. A significant part of animal husbandry in many Indian States *viz.* Rajasthan, Himachal Pradesh, Arunachal Pradesh, Jammu and Kashmir, Andhra Pradesh and Tamil Nadu is still based on nomadic pastoralism. The driving causes behind the tradition of livestock migration are lack of dry-season water resources and acute shortages in fodder. The result is a regional livestock population increasingly on the move  sometimes turning to year-round nomadism to meet demands for seasonal pasturage. Depending on severity of drought, deficit rainfall, temporary migration with lower flock size move by the end of winter/start of summer and the migrants return by the onset of monsoon. Semi-migration with large flock size of animals migrates out of district or State. Permanent migration moves from the home tract to other districts and States and are always on the move.

Animals reared by pastoralists include sheep, goats, cattle, buffaloes, camels, and yaks. While no formal studies have been conducted it is estimated that there are around 35 million pastoralists in India who are rearing 80 per cent of the total population of sheep and to a lesser extent goats. Almost all of camel (0.25 million) and yaks (0.06 million) are reared by pastoralists. Large ruminants *viz.* cattle and buffaloes were also reared under pastoralism but considerable focus on their R&D and breeding for milk and meat production by government, both at the Centre and State levels, as well as various public and private funded developmental agencies over the past several decades has led to their comprehensive growth and sedentarization. Each of the major indigenous breeds – Gir, Sahiwal, Rathi, Tharparkar and Murrah – emerged out of pastoral systems most of which have now been sedentarized. On account of their relentless support and funding by Government agencies and departments India has emerged as the leading milk (188 million tonnes) and buffalo meat (export value US$ 4 billion in 2018) producer in the world. However, while large ruminants have received substantial attention there has been very little focus on small ruminant and other minor ruminant species *viz.* camels and yaks most of which are reared by pastoralists.

Livestock migration can be considered as an essential mechanism for sustainable animal husbandry. The various advantages of migratory pastoralism and the important part played by the pastoralists includes producing organic wealth on the move from zero inputs besides ensuring that the shortage of feed and fodder resources of the country are not put under greater stress. Further, the pastoralists are expert livestock breeders in conserving the country's genetic wealth and making the breeds stress resilient and hardy. If pastoralists are sedentarized, it would be impossible to provide adequate nutrition for all livestock and lead to costly feed and fodder imports besides increasing desertification and hampering efficient ecological sustenance. Hence, keeping the importance of pastoralism in mind, it is necessary for the government to undertake initiatives through a policy advocacy for government schemes to mitigate the challenges being faced by the pastoralists in terms of improving CPRs, developing water bodies along the migratory routes, delineating pastoral rights and welfare of the pastoral communities, introducing digital technology to assist the pastoralists, helping create government assisted marketing channels of their produce among other things. NRAA has initiated a move to prepare necessary guidelines for such a policy which will involve several Ministries.

## Conclusion

Although livestock production as well as productivity is on a growth trajectory, for ensuring zero hunger by 2030 it is imperative that the challenges mentioned have to be taken up in right earnest through well implemented government policies and Centre-State cooperation. It needs to be emphasized here that livestock production cannot be dealt with in isolation. It is intricately linked with crop production, rural livelihood, employment generation, climatic conditions, water resources, land reforms, as well as assessment of risks in livestock production and its management and effective application of new technologies *etc*. A cohesive approach with convergence among various departments of several ministries is necessary for harnessing the synergies and ensuring adequate financial resource management by avoiding duplication through application of a common digital Management Information System platform which will go a long way in providing smooth assistance to the farmers and livestock owners.

# Chapter 17

# Animal Health, a Prerequisite for Better Human Health: One Health Perspective

S.V.S. Malik

*ICAR-IVRI,
Izatnagar, Bareilly, Uttar Pradesh*

Presently we are confronted with the problems of 21st century that include:

(a) Population explosion

(b) Infectious diseases

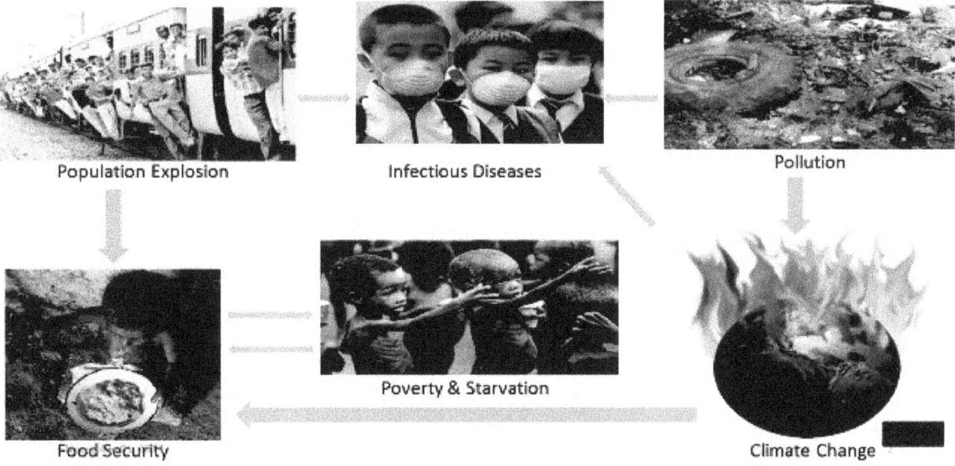

   (c) Pollution

   (d) Food security

   (e) Poverty and starvation and

   (f) Climate change.

Health is physical, social, mental and spiritual well being of a person and his harmony with his environment. It goes true for animals also. Food that sustains the life is coming from the mother earth and both animals and humans do share it and are dependent on it. Since human and animals are interlinked and interdependent, any negligence in the process will lead to dangerous trend, so comes here the food safety. Food safety has emerged as a global issue mainly due to international trade and public health implications. About 250 known diseases do occur as food born diseases and $2/3^{rd}$ of them are caused by bacteria. Food safety is a global issue, especially in developing countries. Some 3.5 billion episodes of diarrhoea, over 1.8 million deaths in children of under-5 year age worldwide in 2010 with significant proportion of it occurring due to consumption of contaminated food with 31 infectious agents (pathogens) or chemicals or toxins (WHO, 2015) are witness to the fact that food safety is of paramount importance in human and animal health. In USA we have the data available of 1997 that shows some 76 million cases annually are affected by food borne diseases caused by major pathogens, ending with 5000 deaths with an estimated cost of US \$35 billion annually. In India some 0.5 million people die annually of diarrhoea. Globally, diarrhoeal infections with *Salmonella* species (including invasive infections), enteropathogenic and enterotoxigenic *Escherichia coli*, norovirus and *Campylobacter* species were responsible for the greatest burden of food borne disease. Food safety measures that would be effective against these enteric pathogens are likely to be similar, at least at the food preparation stage.

Contamination of food can occur anywhere along the chain of production and preparation, be it at the ground level when sown or grown, harvested, processed,

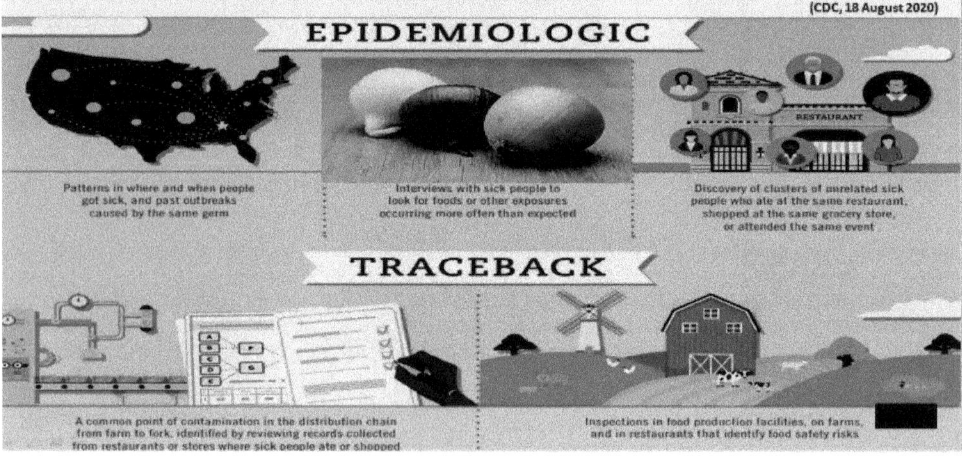

transported and sold, to its preparation before consumption. Thus at all these levels, proper hygiene, storage of food and systemic improvements in food supply chain is particularly important.

## Trade Related Issues

In today's interconnected and interdependent world, international trade, travel and migration has tremendously increased the potential of local food borne disease breakouts to become global threat. There remains always a risk of spread of dangerous pathogens and contaminants in food deliberately or accidentally across the globe. Distribution of contaminated food products affecting the health of people in many countries at the same time period are trade related issues. In 1991 cholera originated from the contaminated sea food harvested from the coast of Peru that spread rapidly across the Latin America and affected 400000 people with over 4000 deaths in many countries. In early 2008 outbreak of avian influenza in Bangalore, India, resulted into the export ban of poultry products to Middle East, ending up with loss of hundred thousands of USD to the Indian economy. One single contaminated food ingredient can lead to the recall of tonns of food products leading to tremendous economic losses in production as well as damage to the industry.

## Food Residues and Adulteration

Pesticides and antibiotic residues in food is a big issue. Indiscriminate use of pesticides and antibiotics has lead to the incorporation and accumulation of these chemicals in the food chain. Overuse of antimicrobials and heavy metals result into antimicrobial resistance, a big challenge now. Similarly adulteration in animal foods has a serious health, religious and ethical issues. This trend sometimes leads to the rejection of consignments particularly to the countries with high food safety standards, resulting in to great economic losses. Various veterolegal issues are being confronted because of food adulteration.

Therefore quality assurance need to be delivered by developing rapid, effective onsite test/assays for product safety, product quality and species of origin of the food. Development of indigenous rapid assay kits and protocols to assure food quality is need of the day. Traceability which is a buzz word in seen only in USA, UK and Australia whereas in India and other developing countries, traceability is not available and biosecurity concerns, environmental contaminants are quite rampant in developed countries. Disease monitoring, recording and reporting is very poor in our conditions and need to be strengthened, data collection risk assessment, average daily intake (ADI) need to be worked out in our settings, Risk based approach is important in our economic standing as we cannot afford to cull/destroy whole lots.

## Chemical Hazards in Foods

Pesticides, heavy metals, drug residues and other chemicals that are in extensive use in India, reach to the soil. Soil health is compromised, reaches into the growing crops and leaches to reach ground water. Pesticide effluents from pesticide industries deposited into the rivers or streams resulting into loss of fauna in these bodies. The health hazards of these chemicals in food are numerous and they can work as endocrine disruptors leading to hormonal imbalance, reproductive failures, they lower the immunity and intelligence and above all increases the risk of cancers. Chemicals like oxytocin, melamine are also used in animals that have serious consequences in animal itself. Cancer causing oil is being used in coating of food items, such residues remain unchecked and both animal and human health is put to high risk of precarious pathologies.

## Antibiotic Residues

Antibiotics in India are widely used in food animals to promote growth, prevent infections and treating infections. India does not have any regulation on controlling antibiotic use or sale in livestock and poultry production industry and is practically open to all.

Of late some regulations have come for use of antibiotics in veterinary and medical practice. It is pathetic to note that honey which we use in infants and elders as elixir do contain high levels of drug residue (Figure 17.1). Poultry meat that serves as an important protein source in the food basket is contaminated with antibiotic residues (Figure 17.2). These antibiotic residues have very serious consequences of drug resistance. Drug withdrawal period is never considered in our Indian conditions; neither we have testing facilities nor checks/regulations to monitor the same.

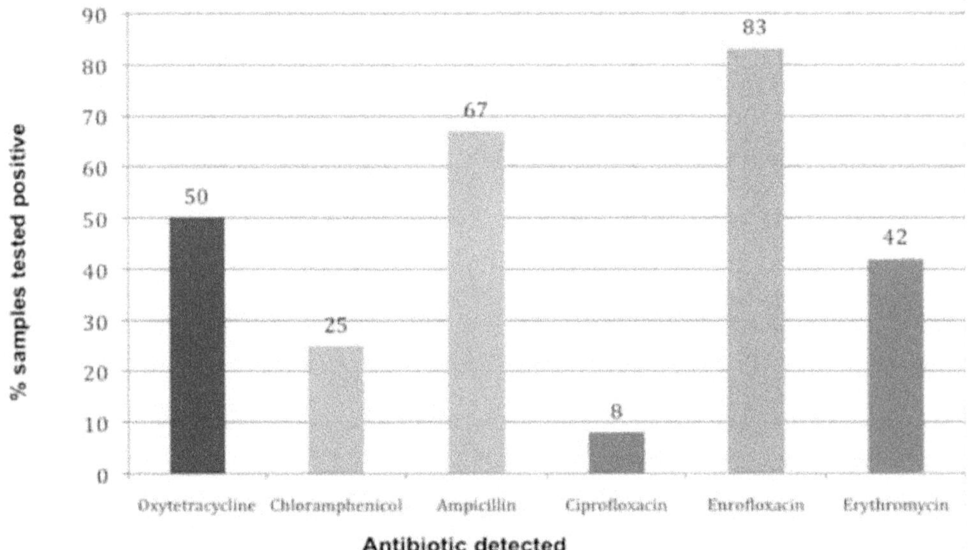


**Figure 17.1: Antibiotic Residues Detected in per cent Honey Samples.**

2/3$^{rd}$ of food poisoning is through bacteria. WHO has recorded worlds largest ever outbreak of Listeria through meat in South Africa with case fatality rate of 20 per cent. The high fatality rate in listeriosis is because, the Listeria enters into the brain and most of the drugs do not cross easily the blood brain barrier. Listeria in western world gets spread through cheese and there are some 200 kinds of cheese. Women particularly are advised not to take cheese due to potential threat of listeria that affects reproductive life besides septicaemia. Outbreaks of Salmonella infection from backyard poultry in USA was reported yet with low CFR of 1 per cent because of being developed country. Recently CDC in August, 2020 has reported Salmonella outbreak linked to peaches. Integrated farming with poultry, fish, and cattle must be carried forward with caution to check spread of diseases across the species.

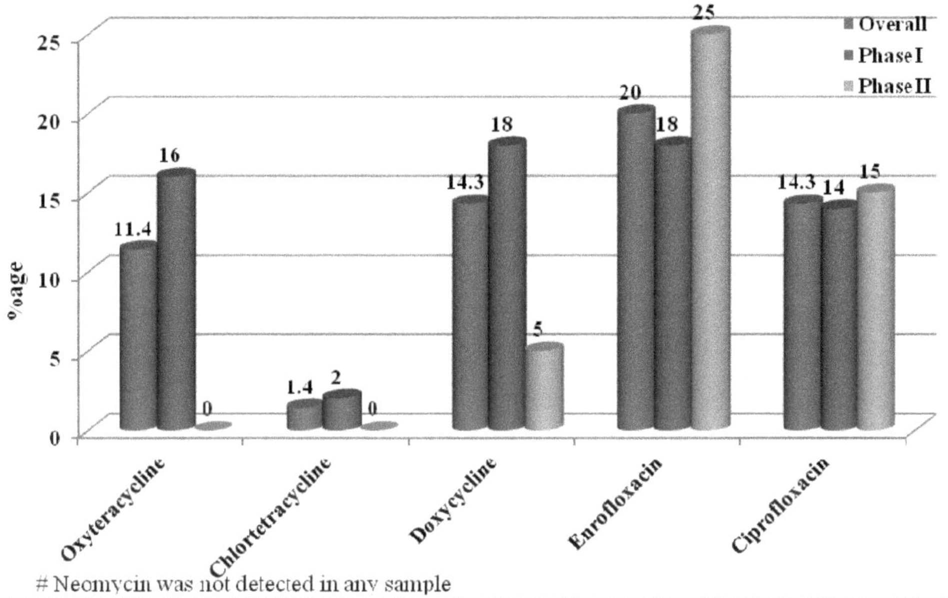


**Figure 17.2**

## Food Safety a Shared Responsibility

It is our shared response to have safe food, get educated about the safe food and share the knowledge and make others aware of it.

### *The responsibility of Government in food safety is*

☆ To maintain up to date and science based legal standards

☆ To ensure effective operation of the national food control system

☆ To verify and provide assurances as to the conformity of food and the associated production with requirements.

☆ To provide health education programmes to effectively communicate the principles of food hygiene to industries and consumers.

### *The responsibilities of food business operators is enlisted as*

☆ It is important for the business operators to realise that only profit is not their priority but they have certain societal responsibilities.

☆ Producing safe food and adhering to the established regulatory requirements.

☆ To withdraw and or recall food from the market if it is not in compliance to the food safety requirements.

☆ Advance and foster knowledge, science, research, education regarding food safety.

### *Consumer responsibilities include*

☆ To read and follow the instructions on product package,

☆ To clean the raw vegetables, hands and food contact surfaces.

☆ To separate raw foods from ready to eat foods

☆ Proper refrigeration of perishable food products

☆ Cooking food to attain proper internal temperature especially when product is coming from different sources where post-processing contamination chances are more.

☆ If in doubtful as to the safety of your food, throw it out, do not let it reach to underprivileged as it will be more catastrophic.

### Everyone has a Role in Food Safety

Unless all the stakeholders be it school administration, team leader, food service staff, school nurses, health departments, cooperative extension, families and students and teachers work together in coordination, food safety cannot be ensured. Too much is talked about food safety, but little is practised, we need to have understand and address the common interest of have safe food.

## Everyone Has a Role in Food Safety

## An Average Indian makes a Morning with Zoonotic Disease

National papers can be seen every other day with the headings like "Zoonotic diseases are rife in this century. Globalized world requires global code of conduct to overcome this threat" (Indian Express, 28-03-2020), "deforestation and disease: how natural habitat destruction can fuel zoonotic diseases" (Indian Express, 01-04-2020), "there is a zoonotic viral disease in South India and it peaks every summer" (The Print, 29-04-2020), "water borne diseases on the rise: here is what to do" (Indian Express, 18-08-2020), "antimicrobial resistance one of the biggest public health threat in the world: NIH Head" Dawn, 21-08-2020).

## Burden of Food Borne Zoonotic Infectious Diseases

Bacterial and viral zoonotic infections comprise a practically endless, ever-expanding list of pathogens that have the potential to induce human disease of varying severity, with varying means of transmission to humans (including vector-borne and food borne agents) and of varying epidemiology. Increasing demand for food due to an expanding global population has led to a substantial susceptibility of our populations to food-borne zoonoses (Newell *et al.*, 2010). Pathogens in the livestock production chain are a particular risk, with repeated outbreaks from meat, eggs, milk, and cheese, or meat byproducts. Globally, most types of domesticated and wild vertebrates and many invertebrates are food for people; such foods are capable of harbouring zoonotic bacteria, viruses, or parasites (Delgado *et al.*, 1999). 1415 diseases are affecting the human out of which 61 per cent (868) are zoonotic in nature that means for every 10 diseases in human, 6 have zoonotic potential. Out of 1415 diseases, 217 are caused by viruses/prions, 538 by bacteria/rickettsia, 307 fungi, 66 protozoa and 287 by helminths (Taylor *et al.*, 2001). Seventy vie per cent of emerging infectious diseases are zoonotic in nature. Five new diseases in human appear every year with 3 coming from animal origin. Emerging diseases like SARS, MERS and now Covid-19 pandemic has dictated the terms to rethink and refocus on the emerging zoonotic diseases, especially with "one health" approach. Realistic assessment of the magnitude, risk posed and impact of emerging and re-emerging zoonoses remains a vital prerequisite for their prioritization and successful management with limited resources.

Food borne pathogens of many zoonotic diseases often cause mild or subclinical disease in reservoir hosts, and because surveillance systems for wildlife and domestic animals are not universally adequate for detection of clinical disease or pathogen presence, human's beings often act as sentinel populations for zoonoses (Rabinowitz *et al.*, 2009). Since it is question of our well being and survival, we need to know and assess the emerging zoonotic diseases so that our limited resources are directed and utilized in best possible way to promote well being. Plague in 14th century has taken away 1/3rd of European population simply because they were not prepared for the same, even today when the development in science has touched the sky, Covid pandemic has shaken whole globe and developed countries in particular due to the failure of health care. In general zoonoses costs heavy to the low income nations with 10 per cent of total DALYs (disability- adjusted life years), whereas it is very less in high income countries only 0.02 per cent (Grace *et al.*, 2012).

More alarming is the trend with more than 300 infectious diseases have emerged since 1940 (Gavi.Org. April 2, 2020) and after every 3 -4 months we witness a new disease, posing a serious threat, a new disease knocks our door when we are still reeling with the previous one (UNEP, 2016).

Wild life is very important source of these emerging diseases, yet we don't have close proximity with wild but fact remains that we have ventured into their habitats. It has been reported that out of 95 zoonotic viruses, 65 per cent were transmitted exclusively from wild animals, 26 per cent from wild as well as domestic animals and 9 per cent exclusively from domestic animals. It has been reported that between 2000 -2013 some 25 emerging and re-emerging infectious disease threats are linked to wild life.

The techniques with which animals are slaughtered and processed, and practices of product storage, packaging, transportation, and preparation at the place they are consumed, are all important links in food borne disease outbreaks. Outbreaks of trichinosis in people are often linked to the consumption of incompletely cooked meat from pigs and wild boars and, occasionally, wild game (Newell *et al.*, 2010). Cysticercosis (caused by the pig tapeworm Taenia solium) affects 50 million people every year. Echinococcosis (caused by the larval stages of the dog tapeworm Echinococcus granulosus for which ungulates serve as the intermediate host) affects 200000 people every year, resulting in relative economic impacts equivalent to US$4 1 billion annually for treatment and control in humans and animals (WHO, 2006). Other notable food borne parasites include trematodes (liver, lung, and intestinal flukes), which are a neglected disease group despite contributing to a substantial disease burden in southeast Asia and posing a serious impediment to public health and economic prosperity in the region (Keiser and Utzinger, 2009).

Religious rituals/congregations have high potential of spreading the pathogens through water, breathing, defecation, poor sanitation *etc.*

## Pathogens do not Respect the Animal Species Boundaries

Zoonotic pathogens reach to our bed rooms. Many important diseases like Plague, Pasteurellosis, Chagas disease, Rabies, Toxocariasis, Giardiasis *etc.* get contracted through sleeping with, kissing and licking of pets. Unfortunately these are not noticed and correctly diagnosed by the medicos. Symptomatic treatment with antibiotics for short duration is posing threat of drug resistance.

## Transmission of Pathogens

The volume of consumption of wildlife products for food is at least an order of magnitude lower than it is for domestic livestock (Karesh and Cook, 2005). However, human being–animal contact associated with hunting, preparation, and consumption of wild animals has led to transmission of notable diseases. Such diseases include HIV/AIDS, which was linked to the butchering of hunted chimpanzees (Hahn *et al.*, 2000). SARS, which emerged in wildlife market and restaurant workers in southern China (Guan *et al.*, 2003) and Ebola haemorrhagic fever linked to the hunting or handling of infected great apes or other wild animals (Rouquet *et al.*, 2005). Many food borne zoonoses are enzootic in livestock (*e.g.*, bovine tuberculosis,

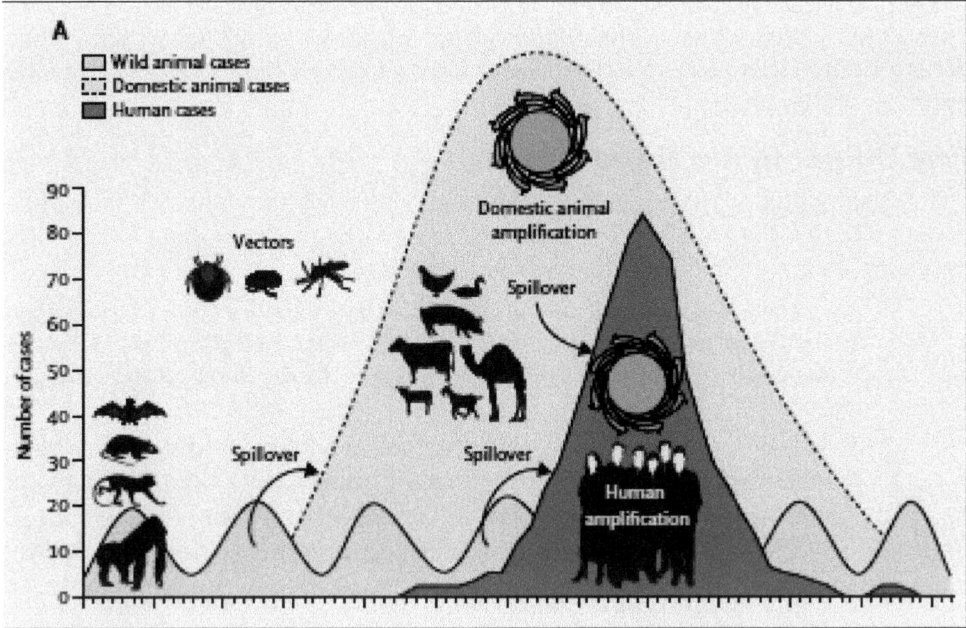


**Figure 17.3**

brucellosis, salmonellosis, and some helminth infections), especially in low-income and middle-income countries, and result in endemic infections and outbreaks of disease in people. Understanding the ecology of zoonotic diseases at the human being–animal interface is a complex challenge. It requires knowledge of animal and human medicine, ecology, sociology, microbial ecology and evolution and the underlying issues that drive increased transmission of pathogens in humans, wildlife and livestock: an idea described as One Health perspective (Karesh and Cook, 2005). Therefore the prevention and response to zoonotic diseases and elimination or mitigation of transmission routes to prevent their emergence will need multi sectoral collaboration. Since zoonoses affect developed and developing countries alike, and readily spreads across national boundaries, mitigation and control of these diseases needs collaboration between ministries of health, environment and agriculture and inter governmental agencies involved in health, trade, food production, and the environment. Understanding of the relation between environmental changes, wildlife population dynamics and the dynamics of their microbes can be used to forecast risk of human infection with enzootic or endemic zoonoses (Figure 17.3). All zoonoses have non-human reservoir hosts, and the dynamics of the pathogen in these hosts often determines the risk of outbreaks in people. This risk can vary with geography, seasons or through multiyear cycles and can depend on factors such as changes in land use, weather, climate, or environment. The figure here depicts the kind of setup in which we do live. Pink zone demarcates the wildlife settings in which all wild life lives along with vectors, ticks, mosquitoes, insects and when they come in contact of domestic animals, pathogens spell-over to them and when pathogens get established in domestic animals and amplify in them, these food animals become reservoirs of pathogens/infectious agents. Lastly when they get multiplied, they spell-over to humans and get established in humans causing outbreak of diseases.

## Five Stages in the Transformation

Transmission of an animal pathogen into a specialized pathogen of humans is depicted in the (Figure 17.4). The progression of pathogen from stage 1 to 5 is not necessary, as many remain stuck at certain stage and may not reach to the stage-5.

1. In stage -1, a microbe that is present in animals but that has not been detected in humans under natural conditions, excluding inadvertently transfer through blood transfusion, organ transplants or hypodermic needles.

2. In stage- 2, pathogen of animals that under natural conditions, has been transmitted from animals to humans ('primary infection') but has not been transmitted between humans ('secondary infection').

3. In stage -3, animal pathogens undergo only a few cycles of secondary transmission between humans, so that occasional human outbreaks triggered by a primary infection soon dies out.

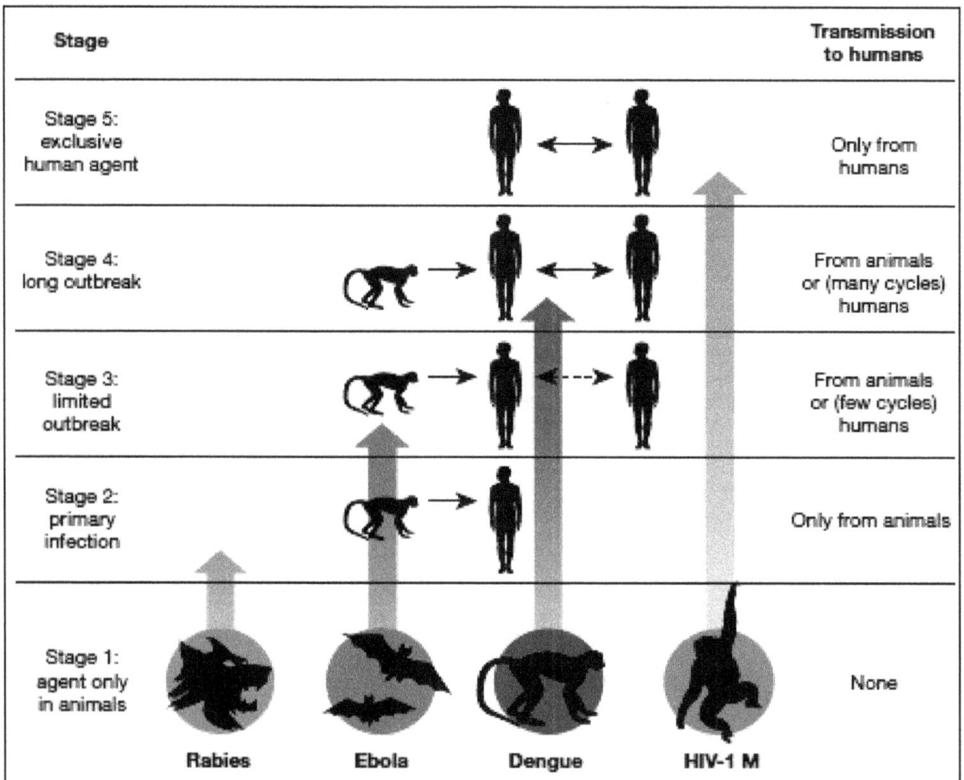


**Figure 17.4**

4. In stage -4, a disease that exists in animals, and that has a natural (sylvatic) cycle of infecting humans by primary transmission from the animal host, but that also undergoes long sequences of secondary transmission between humans without the involvement of animal hosts.

5. In stage -5, a pathogen is exclusive to humans (Wolfe *et al.*, 2007).

## Zoonoses Emergence Linked to Agriculture Intensification and Environmental Change

In nature we do not share only air, land, water, food but also pathogens and toxins.

Agriculture intensification and environmental change are associated with increased risk of zoonotic disease emergence, driven by the impact of expanding human population and changing human behaviour. Human behavioral changes, driven by increasing population, economic and technological development, and the associated spatial expansion of agriculture, are creating novel as well as more intensive interactions between humans, livestock and wildlife. These changes have been implicated as drivers of some recent emerging disease events (Morse, 1995; McMichael, 2004; Woolhouse and Gowtage-Sequeria, 2005), that had important impacts on human livelihoods and health. Sustainable agricultural food systems that minimize the risk of emerging disease will therefore be needed to meet the food requirements of the rising global population, while protecting human health and conserving biodiversity and the environment. The rate of future emergence or re-emergence of zoonotic diseases will be closely linked to the evolution of agriculture-environment nexus.

Encroachment of human settlements and agriculture on natural ecosystems results in expansion of ecotones (transition zones between adjacent ecological systems), where species assemblage from different habitats mix. This provides new opportunities for pathogen spillover, genetic diversification, and adaptation. Associations between disease emergence and ecotones have been suggested for several diseases, including yellow fever, Lyme disease, hantavirus pulmonary syndrome, Nipah virus encephalitis, influenza, rabies, cholera, leptospirosis, malaria, and human African trypanosomiasis (Despommier *et al.*, 2006). Most of these are zoonoses, and several involve both wildlife and livestock in their epidemiology. Geographical expansion of Japanese encephalitis virus (JEV) in Southeast Asia has been associated with increasing irrigated rice production and pig farming due to an expanding human population (Vora, 2008). The combination of irrigated fields, which increase the density of vectors and water birds, and pig farming increases the risk of virus spillover into humans. Intense human settlement and habitat clearance for agriculture has resulted in the disappearance of tsetse flies, which transmit human and animal trypanosomiasis (Ducheyne, *et al.*, 2009). The genetic similarity between *E. coli* isolated from humans and livestock and that of mountain gorillas increased with greater habitat overlap (Rwego *et al.*, 2008). Higher interspecies transmission, which may be in either direction, is therefore likely to arise from greater ecological overlap. Loss of bat habitat due to deforestation and agricultural expansion, changes in the location, size, and structure of bat colonies, and foraging in periurban fruit trees have led to greater contact with livestock and humans, increasing the probability of pathogen spillover (Field, 2009). Water management activities may result in increased density of breeding sites for mosquitoes. Rift Valley fever epidemics have occurred after the construction of dams and irrigation canals (Pepin *et al.*, 2010). Liver fluke and its intermediate snail host have adapted to the irrigation systems of the Nile Delta in Egypt and in Peru, leading to increasing incidence of human fascioliasis (Mas-Coma *et al.*, 2005).

Intensification of livestock production, especially pigs and poultry, facilitates disease transmission by increasing population size and density, such as brucellosis and tuberculosis. Ventilation systems expel material, including pathogens such as *Campylobacter* and avian influenza virus, into the environment, increasing risk of transmission to wild and domestic animals. Large quantities of waste are produced that contain a variety of pathogens capable of survival for several months if left untreated (Perry *et al.*, 2011). Food-borne bacterial pathogens evolve in response to environmental changes, developing new virulence properties and occupying new niches, including antimicrobial resistance facilitated by livestock intensification.

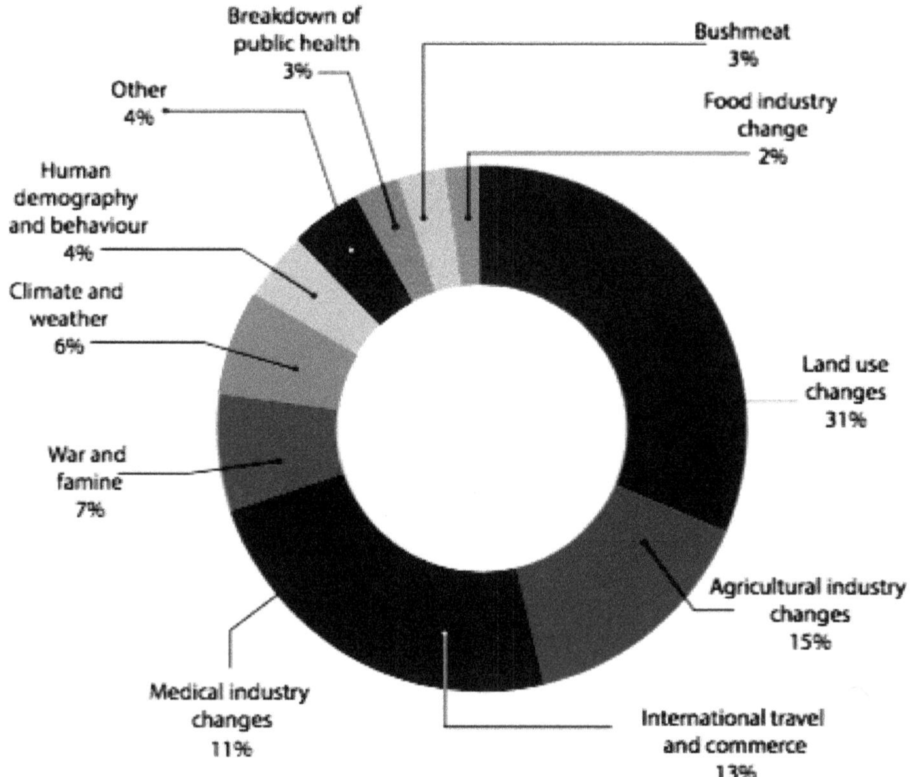


**Figure 17.5**

Loh *et al.* (2015) have shown agriculture activities as the major primary driver of disease emergence. The Figure 17.5 shows that land use and agricultural industry change accounts for 46 per cent of drivers of diseases emergence and re-emergence. International travel and commerce accounts for 13 per cent, medical industry changes- 11 per cent and climate and weather only 6 per cent to the drivers of disease emergence. Therefore, agriculture farming system need to be studied and understood in depth to contain emergence/re-emergence of zoonotic diseases.

Studies are required on the complexity and interrelatedness of environmental, biological, economic and social dimensions of zoonotic pathogen emergence, which significantly limits our ability to predict, prevent and respond to zonootic disease emergence.

# REFERENCES


Delgado C., Rosegrant M., Steinfeld H., Ehui S., Courbois C. 1999. Livestock to 2020 the next food revolution. Food, agriculture, and the environment discussion paper. Washington, DC: International Food Policy Research Institute.

Despommier D., Ellis B., Wilcox B. 2006. The role of ecotones in emerging infectious diseases. EcoHealth **3**(4): 281–289.

Ducheyne E *et al.*, 2009. The impact of habitat fragmentation on tsetse abundance on the plateau of eastern Zambia. Prev Vet Med **91**(1): 11–18.

Field H.E. 2009. Bats and emerging zoonoses: Henipaviruses and SARS. Zoonoses Public Health **56**(6/7): 278–284.

Guan Y., Zheng B.J., He Y.Q., *et al.*, 2003. Isolation and characterization of viruses related to the SARS coronavirus from animals in southvern China. Science. **302:** 276–78.

Hahn B.H., Shaw G.M., De Cock K.M., Sharp P.M. 2000. AIDS as a zoonosis: scientific and public health implications. Science. **287:** 607–14.

Karesh W.B. and Cook R.A.2005.The Human-Animal Link, One world—one health. Foreign Aff. 84: 38–50.

Keiser J. and Utzinger J.2009. Food-borne trematodiases. Clin Microbiol Rev. **22:** 466–83.

Mas-Coma S., Bargues M.D., Valero M.A. 2005. Fascioliasis and other plant-borne trematode zoonoses. Int J Parasitol **35**(11-12): 1255–1278.

McMichael, A.J. 2004. Environmental and social influences on emerging infectious diseases: Past, present and future. Philos Trans R Soc Lond B Biol Sci **359** (1447): 1049–1058.

Morse, S.S. (1995) Factors in the emergence of infectious diseases. Emerg Infect Dis **1**(1): 7–15.

Newell D.G., Koopmans M., Verhoef L., *et al.*, 2010. Food-borne diseases—the challenges of 20 years ago still persist while new ones continue to emerge. Int J Food Microbiol. 139 (suppl 1): S3–15.

Pepin M., Bouloy M., Bird B.H., Kemp A., Paweska J. 2010. Rift Valley fever virus (Bunyaviridae: Phlebovirus): an update on pathogenesis, molecular epidemiology, vectors, diagnostics and prevention. Vet Res **41**(6): 61.

Perry B.D., Grace D., Sones K. 2011. Current drivers and future directions of global livestock disease dynamics. Proc Natl Acad Sci USA doi: 10.1073/pnas.1012953108.

Pfeffer M., Dobler G. 2010. Emergence of zoonotic arboviruses by animal trade and migration. Parasit Vectors **3**(1): 35.

Rabinowitz P., Scotch M., Conti L. 2009. Human and animal sentinels for shared health risks. Vet Ital. 45: 23–24.

Rouquet P., Froment J.M., Bermejo M., *et al.*, 2005. Wild animal Mortality monitoring and human ebola outbreaks, Gabon and Republic of Congo, 2001–2003. Emerg Infect Dis. 11: 283–90.

Rwego I.B.,IsabiryeBasuta G., Gillespie T.R., Goldberg T.L. 2008. Gastrointestinal bacterial transmission among humans, mountain gorillas, and livestock in Bwindi Impenetrable National Park, Uganda. Conserv Biol **22**(6): 1600–1607.

Taylor L.H., S. M Latham, M. E. Woolhouse 2001. Risk factors for human disease emergence Phil Trans R Soc Lond B Biol Sci, 356. pp. 983-989

UNEP. 2016. UNEP Frontiers 2016 Report: Emerging Issues of Environmental Concern. United Nations Environment Programme, Nairobi.

Vora N. 200. Impact of anthropogenic environmental alterations on vector-borne diseases. Medscape J Med **10**(10): 238–245.

Wolfe N.D., Dunavan C.P., Diamond J. 2007.Origins of major human infectious diseases. Nature 2007; 447: 279–83.

Woolhouse, M.E.J., Gowtage-Sequeria, S. 2005. Host range and emerging and reemerging pathogens. Emerg Infect Dis. **11**(12): 1842–1847.

World Health Organization. 2015. WHO estimates of the global burden of food borne diseases. Food borne diseases burden epidemiology reference group 2007–2015. Geneva.

# Chapter 18

# Dairy Food Production Interventions for Livelihood Support and Food and Nutritional Security

**Mohammad Ashraf Paul**

*Professor and Dean Faculty of Veterinary Sciences
and Animal Husbandry, SKUAST-Kashmir*

Dairy Food Products are the food products of animal origin consumed by human beings for sustenance. These supplement vegetable proteins in modern balanced dietary concept and are classified as protective foods. They have an excellent nutritional value and unique sensorial characteristics that can hardly be mimicked by any available artificial sources.

## Objectives

- ☆ Better utilization and value addition of the produce.
- ☆ Minimizing wastage at all stages in the food processing chain by using appropriate storage, transportation and processing.
- ☆ Introduction of modern concepts of quality assurance.
- ☆ Encouraging R&D in food processing for the innovative products and process development.
- ☆ Build necessary infrastructure to fill the gaps in the supply chain from farm to fork.
- ☆ Image building via adherence to TQM.

# Significance

- ☆ Efficiency of animal production is largely dependent on efficient processing and utilization of the produce.
- ☆ A correlation exists between the livestock products consumption and socio-economic status.
- ☆ With the current trend of increasing socio-economic status the consumption is anticipipated to increase significantly.
- ☆ Aptest way of minimizing losses occurring due to regional and seasonal imbalances.
- ☆ Nutritional and food security.
- ☆ Immunity enhancement to withstand pandemics.
- ☆ Alleviation of the problems of mal-nutrition.
- ☆ Alleviation of the problem of hidden hunger.
- ☆ Employment generation.
- ☆ Poverty alleviation.

### Dairy Production Standing

| Characteristics | India | J&K |
|---|---|---|
| Milk production (mMT) | 187 | 2.541 |
| PCA (g/day) | 375 | 401 |
| DCS | 144246 | — |
| Farmer members | 14,461,000 | — |
| Processing cap. (LLPD | 300 | ~2.0 |
| FSSAI Reg. plants | 678 | 05 |
| Exports ($ billion,2018-19) | 28.6 | — |
| Consumption value/person/month (share in diet, per cent food) | Rural: Rs.116.4 (18.7<br>Urban: Rs.187.1 (20.3) | — |

# Prospective Interventions

1. Dairy Cooperative Societies
2. Cottage Dairy Industry (CDI) concept
3. Village level chilling centres
4. Bactofugation units
5. Organic products
6. Equipment and ingredients manufacture
7. Value added/Functional product manufacture
8. Cheese production units
9. Self help group concept
10. Dairy Food Parks

# Dairy Cooperative Societies

## Village Level Cooperative Dairy Society

- ☆ Primary Producer basic member.
- ☆ Elect 9- member management committee with a chairman amongst them.
- ☆ Main Function - Collection, testing, payment.

## District Level Milk Union

Village Cooperative Dairy Societies confederate to form District level milk union.

- ☆ Main Function-Storage.

## State Level Milk Federation

- ☆ District level milk unions Unite to form SLMF.
- ☆ Main job- procurement, transport, processing and marketing.

## Cottage Dairy Industry (CDI) Concept

It is an occupation of farmers, generally landless or with very small cultivable land that cannot be used for raising animals for commercial milk production. These farmers keep 3-5 animals for milk, meat and power source requirements.

## Salient Features

- ☆ Potential to alleviate poverty.
- ☆ Nutritional security.
- ☆ Promising employment opportunities.
- ☆ Provides income on a daily basis.
- ☆ Provides salubrious milk for consumption.
- ☆ Hunger mitigation.

## Main Benefits

- ☆ Supplements agriculture being more dependable.
- ☆ Provides perennial source of income contrary to crop production.
- ☆ Work force in animal management increased compared to crop production.
- ☆ Contribution of livestock to GDP increased from 4.8 to 5.5 per cent in last 2 decades.
- ☆ Unemployed family members (women) get employed.
- ☆ Money earned used for upkeep of animals as well as family.
- ☆ Judicious land use.

## Challenges

- ☆ Milk production per animal is low.

- ☆ Farmers mostly with small land holdings.
- ☆ Milk producers are clustered in a village, Villages scattered and away from urban consumption centers.
- ☆ Organization of this scatter to weave a network is required to make the system work as an industry.
- ☆ Regular payment to producers.
- ☆ Assured veterinary health care and insurance.
- ☆ Building infrastructure-roads and telecom.
- ☆ Networking Veterinarians in emergencies.
- ☆ Assured marketing backup.

| Animal Health Care | Cattle Feed | AI | Animal Management | Education |
|---|---|---|---|---|
| Package of services for milk production enhancement | | | | |

## Bactofugation Units

- ☆ Centrifugal process employing two centrifugal clarifiers in a series I$^{st}$ operating @ 20,000 rpm. with the objective of removing maximum number of micro-organisms.
- ☆ 99 per cent of bacteria are removed.
- ☆ Triples shelf life of market milk.
- ☆ Supplementary but not an alternative to pasteurization.

## A Representative Village Level Chilling Centre is Illustrated

RKVY funding is being provided via government agencies for the establishment of Milk chilling centers in rural production areas for central processing in urban consumption areas.

**Organizational Model for CDI**

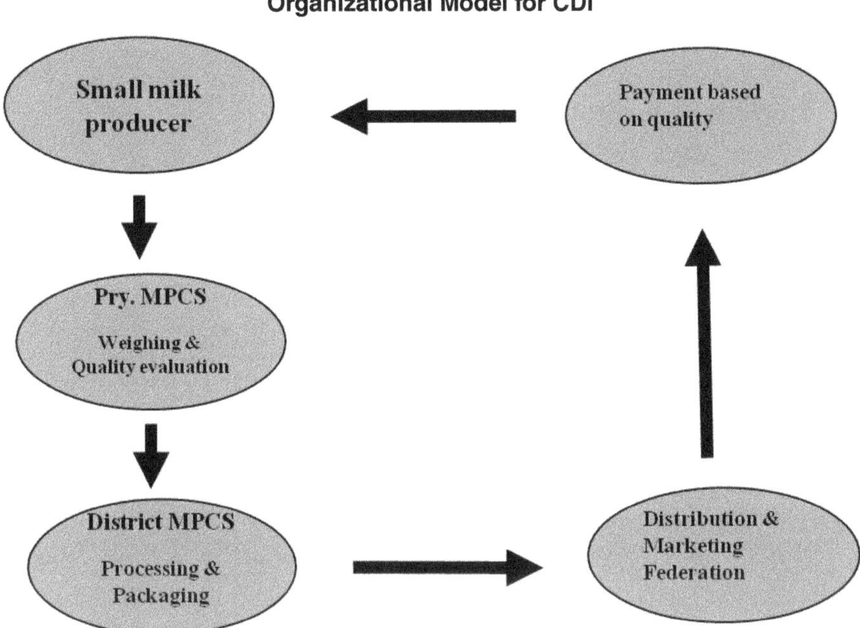


# Organic Dairy Products

## Favourable Features

- ✮ Existence of mixed farming systems.
- ✮ Hardy and disease resistant native dairy breeds.
- ✮ A significant area that can be converted to organics.
- ✮ Cheap and easily available labour.
- ✮ Rainfed agriculture.
- ✮ Increasing health consciousness *etc*.
- ✮ Promising market/Expanding clientele.
- ✮ Increasing annual growth.

## Equipment and Ingredients Manufacture

- ✮ Processing plants.
- ✮ Accessories.
- ✮ Ingredients.
- ✮ Consumables and spares.
- ✮ Refrigerated transport.
- ✮ Starter culture propagation units.
- ✮ Packaging materials.

☆ Printing, graphics and barcode designing.

☆ Advertising.

## Value Added Product Manufacture

☆ Sweet meats-ready mix.

☆ Functional products.

☆ Improvised Indigenous products *e.g.* Processed Paneer.

☆ Byproduct utilization *e.g.* ghee residue, whey, lactose *etc.*

## Cheese Production Units

### *For use in*

☆ Metropolitan cities.

☆ Hotels.

☆ Restaurants.

☆ Cafés.

☆ Special hostels.

☆ Hospitals functions.

☆ Feasts and Functions.

## Self-Help Group Concept

Association of Socio-economically homogenous people with a background of affinity who meet regularly to transact business with the objective of furthering their common interests for progress and prosperity

☆ An ideal group comprises 15-20 members.

☆ Same Socio-economic status with strong affinity.

☆ Rotational leadership.

☆ Voluntary participation.

☆ Social viability.

☆ Non-partisan democratic decision making.

☆ Rules and regulations framed and followed.

☆ Accounts maintained and updated.

☆ Facilitating Institutions: NABARD, MYRADA *etc.*

## Dairy Food Parks

☆ Centralized Operative system housing various sections of milk and Milk products processing units at one place.

☆ Transformation of Dairy farming into a High tech industry.

☆ Farmers/SHGs process their produce on cost basis during specified hours.

## Objectives

- ☆ Employment for youth.
- ☆ Effective functioning of farmers,SHGs, NGOs *etc.* through utilization of revolving fund.
- ☆ Availability of milk and milk products at affordable prices to rural population.
- ☆ Proper utilization of milk without spoilage.
- ☆ Remunerative price policy.
- ☆ Timely payment.
- ☆ Livestock support services.
- ☆ Quality livestock feed availability at reasonable prices.
- ☆ Risk management.
- ☆ Periodical training programmes – Milk processing, Value addition, hygienic milk production, AI, GMP, HACCP, SOP *etc.*
- ☆ Introduction of community milking.
- ☆ Introduction of central cleaning machinery for equipments *etc.*
- ☆ Effective management of CPRs.
- ☆ Overall enhanced productivity.

## Economic Viability Chart of Dairy Food Parks

- ☆ Procurement prices (Fat/SNF, SCC).
- ☆ Cost of production (recurring, depreciation, wages *etc.*).
- ☆ Price Fixation.
- ☆ (competitive/conforming to highest quality standards).
- ☆ Cost of production (70 per cent); DFP(15 per cent) ;Distributors(5 per cent); Dealers (10 per cent).
- ☆ Payment to SHGs, NGOs *etc.*
- ☆ Issue of debit cards for cash withdrawal.
- ☆ Net Profit.
- ☆ Producers (90 per cent); DFP (5 per cent); SHGs/NGO (5 per cent).

# Chapter 19

# Aquaculture: Mitigating Global Hunger

**M.H. Balkhi, Anayit ullah Chesti and Mansoor Rather**

*Faculty of Fisheries, SKUAST-K, Rangil, Ganderbal*

Fish "Rich food for Poor"; cheapest animal protein. Fish provides more than over 20 per cent of animal protein to 2.6 billion people globally. Fish's unique nutritional properties make it also essential to the health of billions of consumers in both developed and developing countries. Fish is one of the most efficient converters of feed into high quality food and its carbon footprint is lower compared to other animal production systems. Through fish-related activities (fisheries and aquaculture but also processing and trading), fish contribute substantially to the income and therefore to the indirect food security of more than 10 per cent of the world population, essentially in developing and emergent countries. Fish is critically important to food security and good nutrition. Fish and other aquatic foods are high in protein and contain many essential micronutrients. The fishery and aquaculture sectors are the source of income for millions of women and men in low income families, thus contributing directly and indirectly to their food security.

Traditionally the major focus of nutritionists was on the macronutrients providing energy and protein. Increasingly today the role of micronutrients-vitamins and minerals – in the diet, particularly of the poor, is recognized as having a limiting effect on development and health. Micronutrient deficiencies affect hundreds of million people, particularly women and children in the developing world. More than 250 million children worldwide are at risk of vitamin A deficiency, 200 million people have goitre and 20 million are mentally retarded as a result of iodine deficiency, 2 billion people (over 30 per cent of the world's population) are iron deficient, and 800 000 child deaths per year are attributable to zinc deficiency. Rural diets in many countries are not particularly diverse, and thus, it is vital to have a good food sources that can provide all essential nutrients to their diets. Foods from the aquatic environment are a complete and unique source of both the macro- and micronutrients required in a healthy diet. The experts found convincing evidence of beneficial health outcomes from fish consumption for: reduction in the risk of death from coronary heart disease and improved neuro-development in infants and young children when the mother consumes fish before and during pregnancy

Long-chain omega-3 fatty acid docosahexaenoic acid (DHA) is important for optimal brain and neurodevelopment in children and eicosapantaenoic acid (EPA) that improves cardio-vascular health. Although many vegetable oils contain omega-3 fatty acids this is in the form of alpha-linolenic acid (ALA), which must be converted metabolically by chain length extension to EPA and DHA. However, the conversion from ALA into EPA and DHA is limited in humans, making it difficult to rely only on vegetable oil during the most critical periods of life. It has been demonstrated that the metabolic pathway in the human male is only 5 per cent efficient, although the rate in females is higher, indicating a bigger requirement. Omega-3 fatty acids in the form of DHA rather than ALA are therefore needed to secure an optimal brain and neural system development in neonates and infants. This is particularly important during pregnancy and the first two years of life.

The latest estimate suggests for instance that, fish accounted for 17 per cent of the global population's intake of animal protein and 6.5 per cent of all protein consumed. Fish is more than just a source of animal protein. Fish contains several essential amino acids, especially lysine and methionine. The lipid composition of fish, with the presence of long-chain, poly-unsaturated fatty acids (LC-PUFAs), is unique. Fish is also an important source of essential micronutrients – vitamins D, A and B, and minerals (calcium, phosphorus, iodine, zinc, iron and selenium), which makes it particularly attractive in the current fight against malnutrition in low income and food deficient countries (LIFDCs). Complementing its fatty acid content, fish is also known to be an important source of essential micronutrients – vitamins D, B and minerals. Lipid-rich fish also contain vitamin A. In addition to the health benefits of these macro-nutrients fish is also an important provider of a range of micro-nutrients, not widely available from other sources in the diets of the poor. More and more attention is given to fisheries products as a source of micronutrients such as vitamins and minerals. In addition, fish enhances the bio-availability of iron and zinc from the other foods in a meal. Wild and farmed fish are a healthier alternative to almost any other meats. Farmed fish have a more constant nutrient composition compared to their wild counterparts, whose environment, food and access to food varies during the year. The environment of farmed fish can be monitored and managed to secure an optimal product. By controlling the composition of aquaculture feeds and other inputs, fish with good health and healthy fish products with optimal nutritional composition can be produced.

Global fish production is estimated to have reached about 179 million tonnes in 2018. Of the overall total, 156 million tonnes were used for human consumption, equivalent to an estimated annual supply of 20.5 kg per capita. Aquaculture accounted for 46 per cent of the total production and 52 per cent of fish for human consumption. World aquaculture production attained another all-time record high of 114.5 million tonnes in live weight in 2018, with a total sale value of USD 263.6 billion. World aquaculture production of farmed aquatic animals grew on average at 5.3 per cent per year in the period 2001–2018. About 88 per cent of the 179 million tonnes of total fish production was utilized for direct human consumption, while the remaining 12 per cent was used for non-food purposes. In 2018, live, fresh or chilled fish still represented the largest share of fish utilized for direct human consumption (44 per cent). In the period 1961–2017, the average annual growth rate of total food fish consumption increased at 3.1 percent, outpacing annual population growth rate (1.6 per cent). In per capita terms, food fish consumption rose from 9.0 kg (live weight equivalent) in 1961 to 20.3 kg in 2017. Preliminary estimates for per capita fish consumption in 2018 currently stand at 20.5 kg. The expansion in consumption has been driven not only by increases in production, but also by a combination of many other factors: technological developments; rising incomes worldwide; reductions in loss and waste; and increased awareness of the health benefits of fish. Over 90 per cent of the quantity (live weight equivalent) of trade in fish and fish products consisted of processed products (*i.e.* excluding live and fresh whole fish) in 2018, with frozen products representing the highest share. About 78 per cent of the quantity exported consisted of products destined for human consumption.

Globally, aquaculture provides about half of all fish for human consumption. And with half of all wild fish stocks now harvested to full capacity and a quarter over-exploited, we can expect aquaculture's share of fish production to increase further. This can benefit poor people by improving their food security and nutrition, creating jobs, stimulating economic growth and offering greater diversification of their livelihoods. Although we cannot greatly increase catches from capture fisheries, wild fish stocks remain vital to many national economies and to the day-to-day welfare of millions of people. So it is essential that we sustain current catches and grasp opportunities to use the fish we catch better and add to their value. Failure to sustain and make the most of the catch will have profound consequences for the health, income, livelihoods and well-being of poor people in many developing countries. Aquaculture is one of the fastest growing agriculture commodity with annual growth of >6 per cent in last two decades increased from <14.9 million tons in 1985 to 82.1 million tons in 2018 80 per cent comes from 20 million small-holder farms (<2ha) in developing countries. Also the environmental demands for per unit biomass or protein produced are lower as compared to poultry, piggery and beef (Grain needed for production of 1kg protein of fish: <13kg; pork 38kg; beef 61kg). With the breakthroughs in research and development domestication of more species 336 species are available for farming purposes.

Although we cannot greatly increase catches from capture fisheries, however, it is essential that we sustain current catches. Failure to sustain and make the most of the catch will have profound consequences for the health, income, livelihoods and well-being of poor people in many developing countries. So the future is the enhancement in fisheries and aquaculture production through, "development through resilient 'small - scale fisheries" and "sustainable aquaculture". So, more efforts should be focussed towards, the adoption of sustainable aquaculture that benefit the poor and also in making small-scale fisheries more resilient and productive. Recognizing the complexity and the scale of the challenges faced, the roadmaps to assess where to focus needs to be created by providing knowledge, tools, models or other products that can help most to bring about needed changes and improvements. This has lead the FAO, to devise the vision 2030, in order to boost overall fish production from wild and the culture sources, in which total fish production is expected to expand from 179 million tonnes in 2018 to 204 million tonnes in 2030 and the aquaculture production is projected to reach 109 million tonnes in 2030, an increase of 32 per cent (26 million tonnes) over 2018. Asia will continue to dominate the aquaculture sector and will be responsible for more than 89 per cent of the increase in production by 2030. The sector is expected to expand most in Africa (up 48 per cent), driven by the additional culturing capacity put in place in recent years. The majority (62 per cent) of global aquaculture production in 2030 will be composed of freshwater species, such as carp and Pangas catfish. Also the production of higher-value species, such as shrimps, salmon and trout, is also projected to continue to grow. Similarly, the capture fisheries, Capture fisheries production is projected to stay at high levels, reaching about 96 million tonnes in 2030.

In per capita terms, world fish consumption is projected to reach 21.5 kg in 2030, up from 20.5 kg in 2018. In 2030, about 59 per cent of the fish available for human consumption is expected to originate from aquaculture production. The bulk of the growth in fish exports is projected to originate from Asia, which will account for about 73 per cent of the additional exported volumes by 2030. World trade in fish for human consumption is expected to grow by 9 per cent by 2030. In recent years capture fishery production has been flat, at around 90 million tonnes per year, while aquaculture has continued to show sustained growth – currently around 6.5 per cent a year - faster than all other food sectors. Some gains in capture fisheries might be possible by adopting better management through an eco-system approach, but significant increases are unlikely. However, it has been estimated that if all inputs were available, aquaculture could provide 16 – 47 million additional tonnes of fish by 2030 (Hall *et al.*, 2013). However, there are certain challenges that are faced by the sector which need an immediate mitigation measures to reassure the growth and production, which include: Less utilization and low productivity of water bodies, growing demands for quality seed, feed and fertilizers, increasingly competition with other resource users, lack of proper policy to lease public water bodies, lack of processing facilities and value addition, assurance of food safety and quality - export potential, research, training and capacity development, germplasm exchange, alternative livelihoods support to fishers during non-fishing and banned fishing seasons and security of fishermen along the maritime boundaries. In order to overcome such challenges, various steps have been considered from time to time by different governments, policy makers and organizations, to coordinate fishery development with an integrated and holistic approaches like Integrated development of crop, horticulture, forestry, fishery, poultry, animal husbandry and food processing industry, Utilize neglected water bodies, aquaculture in seasonal water bodies, establishment of Brood Banks and Seed Certifications, nutrient profiling - to list fish as health food (which may include in National Food Security Mission), use of farm-made fish feeds, genetics stocks assessment, disease surveillance and monitoring, aquaculture diversification and water budgeting, encouraging fishermen to use smart technology- Cage culture and GP, allowing FDI in deep sea fishing, Germplasm exchange and institutional linkages and capacity building. Above all, Making fisheries and aquaculture more adaptive to climate changes Analyzing vulnerabilities of fishery and aquaculture systems. Bringing unutilized water bodies in to aquaculture as an horizontal expansion of aquaculture for sustainability and practicing aquaculture in seasonal water bodies as climate change impact mitigation strategy are to be considered to achieve an enhanced fish production with sustainability.

Chapter 20

# Achieving Zero Hunger Target: Innovative Solutions through Science and Technology

**Sheikh Firdous Ahmad**

*Scientist, ICAR- NRC Pig,*
*Rani, Guwhati, Assam*

## ABSTRACT

*India is globally recognized as one of the twelve mega-biodiverse countries and agriculture is a major sector with around 15 per cent contribution to the total national gross domestic product (GDP). It is a sector which provides employment to major portion of countries workforce. A lot has been done in this field in the past, such as Green (crop), White (milk) and Silver (egg/poultry) revolutions, which has greatly boosted the agriculture production; a lot needs to be done to meet the demands of increasing population. The author has tried to put his viewpoints, which are backed by scientific principles. Use of innovative technologies such as Artificial intelligence, Internet of Things and machine learning, geographical information system, CRISPR-Cas system-based biotechnological intervention, Apeel science, cold sterilization, use of Bio-peptides and water retaining superabsorbent polymers. Similarly, innovative use of smartphone technology possesses the potential to produce a huge impact on the agricultural economy in Indian scenario.*

## Introduction

Agriculture sector forms an integral component of the Indian Economy. In India, agriculture and allied sectors form one of the largest sources of livelihood with more than half of the Indian population dependent on it, directly or indirectly. It contributes around 15 per cent contribution to the total national gross domestic product (GDP), though the contribution is fluctuating. As per the Government's economic survey report, "Indian agriculture in one way is a victim of its own past success – especially the green revolution". A virtual complacency after green revolution has resulted in stagnant growth of Indian agriculture in the recent few decades. With increasing population and changing agro-climatic trends, a strong need is felt now to transform the Agrarian economy with innovative scientific and technological interventions.

The food production from Agriculture, allied sectors in India is sufficient as per the available data. This is evident from production and export data related to rice, wheat, cereals, pulses, milk and other food ingredients. However, Indian nation is still home to 25 per cent of malnourished and undernourished world population. At present, the major issue confronting India is uneven distribution. Furthermore, with population explosion, changing agro-climatic trends and varied feed habits, food production will be a challenge by itself in near future. The food production needs to increase by around 70-75 per cent to feed the population by the year 2050. Equalizing distribution along with innovative scientific and technological solutions to improve food production is the key forward. Agriculture and allied sectors hold the key for improving the food production status; however, we need a transformation of agrarian economy so that future challenges are faced meticulously.

Before any planning or attempts of transformation; an efficient strength, weaknesses, opportunities and threats (SWOT) analysis is required. The table below summarizes the SWOT review for Indian agriculture:

**Table 20.1: The SWOT Review for Indian Agriculture**

| Strengths | Weaknesses | Opportunities | Threats |
|---|---|---|---|
| Vast and diverse resources | Low Productivity | Agriculture- main enterprise | Non-sustaining investment |
| Mega-biodiversity region | Heavy dependency on monsoon | Irrigation facilities | Climate change |
| Diverse agro-climatic zones | Small land-holdings and livestock rearing | Improved Seed/ germplasm | High fluctuationsin prices and production |
| Huge livestock inventory (both in number and diversity) | Weak market-, storage- and transport- infrastructure | Precision farming | Undeveloped markets |
| World's second largest irrigation system | Little value addition | Bio-inputs | Small land-holding |
| Abundant skilful manpower | Threat of climate change | Market linkages | International competitiveness |

The following innovative scientific and technological interventions, with a meticulous and dedicated application, can greatly help in transforming the agrarian economy in India:

## 1. Artificial Intelligence (AI), Internet of Things and Machine Learning

AI, Internet of Things (IoT), robotics and artificial neural networking (ANN) are the best innovative intervention in agriculture. Artificial intelligence (AI) has a great role to play in transforming and has the potential to contribute to almost every field of agriculture. In the next few years, a farmer will be equipped with forecasting skills of the highest grade. Forecasting of crop size, early grading of agricultural produce and prompt prediction of disease outbreaks in crops and animals. This may revolutionize the agrarian economy. These will be based on image prediction algorithms with the support of AI, IoT, ANN, machine learning and other such technologies.

These improved technologies offer possibility of efficient water monitoring solutions besides ensuring better productivity by reducing water wastage. It involves the use of AI- and IoT-based wireless soil sensors and real-time analytics. Based on results from AI sensors, the soil moisture and mineral levels may be efficiently calculated, thus facilitating precision agriculture farming. Besides AI, remote sensing technology and global positioning satellites are helpful in assessing soil moisture levels and thus efficient irrigation and fertilizer/pesticide application may be practiced. AI-based fruit picker and crop harvester will be a realization very soon. Precision agriculture promotes resource use efficiency which shall be helpful in two main ways *i.e.*, reducing the input costs with efficient production and ensuring environmental stability and sustainability.

## 2. Improved Precision Farming Approach

Uneven distribution of resources creates a big hindrance for the continuous progress of Indian agriculture. One agro-climatic zone may be adequately irrigated, other be short of irrigation. One zone may be blessed with adequate micro-minerals; however, deficient in another nutrient. Therefore, the efficient and precise understanding regarding the make-up of our land resource(s) is needed. It has led us to the practice of 'precision farming'. In precision farming, different farming parameters, including soil monitoring, weather monitoring and air monitoring are used to generate data that reaches the main database server. Real-time analysis of this data will be helpful in making informed decisions regarding farming operations. This shall help in making early and prompt decisions for better farming. Though precision farming is gaining momentum in India, its application needs high tech innovative interventions. Global positioning system (GPS), geographic information system (GIS), remote sensing, microwave and thermal sensing, geospatial analytics, AI and IoT based studies, ANN-based analysis, fertilizer deep placement and big data analysis are some of the innovative steps needed to make precision farming a great success. Based on data from the above technologies, we shall make precision farming more intense that will subsequently help in better and sustainable utilization of available resources. For small farmers, use of robotics, drones, sensors and radio imaging technology may be very helpful in Indian conditions. However, we need integrated and cooperative approach to make the use of technologies fruitful. In livestock farming, the approach of feeding micro-and nano-nutrients can be applied for realization of the optimum production potential of animals. Overall, intense precision farming is aimed at optimal use of resources, minimum waste and sustainable agriculture.

## 3.  Intense Application of Ionomics

Ionomics is the study of all the mineral nutrients and trace elements found in an organism. Applied ionomics-based are aimed at studying the changes in the ionome in response to physiological stimuli, developmental state and genetic modifications (Salt *et al.*, 2008). This emerging field has wide applications in agriculture and allied sectors for various purposes including covering all the variation in a crop field or an animal farm. Extending ionomics to agriculture and allied sectors (mainly livestock sector) is need of the hour. It promises to help in functional genomics, gene network analysis along with disease diagnosis and therapeutics. In combination with different *omics*-based approaches (genomics, phenomics, transcriptomics and metabolomics), ionomics offers great potential to better understand the relation of genotype with phenotype and design appropriate interventions for better production and reproduction. Ionomics is set to play a significant role in improving traits of crop yield and abiotic stress tolerance, enhancing resistance to pests and diseases and improved product quality (Singh *et al.*, 2015). Ionomics has recently helped in the understanding of patho-physiology of milk fever (post-parturient paraplegia) in animals and has opened new avenues for development of bio-markers and prevention of this disease (Zhang *et al.*, 2018). Ionomics, when applied on similar lines in other fields, has the ability to transform agrarian economy so that future challenges are met meticulously.

## 4.  CRISPR-Cas System-Based Biotechnological Intervention

The gains through traditional breeding are slow and have ceiling effects associated with them *i.e.*, they cannot go beyond certain limits. Different biotechnological interventions (from rDNA technology to transgenesis) started developing around 1970s and 80s that aimed at improved gains from agriculture and allied sectors. The micro-scissors of clustered regularly interspaced short palindromic repeats (CRISPR) offer great potential to realize tangible and rapid changes in economic traits of any living organism from viruses to humans. Agriculture and allied sectors are no exception to this. CRISPR presents a great chance to modify existing characteristics or incorporate novel traits into an organism within a short period of time (Zhang and Zhou, 2014). By introducing small changes in the promoter region of different genes controlling qualitative or quantitative traits, considerable genetic improvements are attainable. Quantitative traits in different crops such as *locule number* (fruit shape and size), *fascinated* (large fruit size), *compound inflorescence* (flower proliferation) and *self-pruning* (flowering time) have been targeted by different researchers (Rodriguez-Leal *et al.*, 2017). Researchers have also generated several new alleles with improved traits like fruit shape, size as well as the architecture of a plant. Genes that negatively regulate the economic traits can be silenced using CRISPR technology with considerable success rates. Xu *et al.* (2016) mutated three genes (*GW2, GW5, TGW6*) simultaneously in rice and were successful in realizing up to 30 per cent increase in seed size. Higher seed number in crops and increased muscle mass in animals are possible by introducing CRISPR-based mutations in *CLVTA3* and myostatin genes, respectively (Haque *et al.*, 2018). Improved quality attributes of agricultural produce is another important target with CRISPR-based micro-scissor technology.

Abiotic stresses are major hindrances for efficient production in agriculture and livestock. The threat of global warming is looming large on agriculture and allied sectors. However, variation does occur in our crops and livestock with regard to traits of tolerance to abiotic stress and disease incidence. For instance, ethylene - an important phytohormone, is responsible for increased tolerance of plants to various abiotic stresses including high temperature and drought-like conditions (Kawakami *et al.*, 2010). Similarly, our indigenous livestock is tolerant to several important diseases like foot and mouth disease, tick infestation, piroplasmosis and other stressful conditions. Subsequent to efficient understating the underlying basis of these tolerance related traits, they can be efficiently exploited using CRISPR technology for overall animal and human welfare. With poultry population of 729 million, if the genetic pathway of egg is intervened, using CRISPR, in such a way that a hen produces an egg per 23 hours instead of current 24-26 hours, the agrarian economy shall transform within shortest possible time. Same is the case with other production quantitative traits in agriculture and allied sectors. Therefore, we propose the modification of quantitative polygenes using CRISPR.

## 5. Innovative Interventions Aimed at Controlling Post-harvest Losses

A whopping amount, amounting around 1 lac crore, is lost as harvest and post-harvest losses of major agricultural produce in India (PIB, GoI). According to an estimate 40 per cent of stored food resources are rendered inconsumable (to humans) due to one or other reasons. The infrastructure and scientific support for preventing these losses is poor in India. Besides providing adequate infrastructure, the following interventions promise significant help to considerably reduce these losses:

### a. Apeel Science

Preservation of food requires a natural solution and not the use of synthetic chemicals. Apeel is a set of natural ingredients obtained from plants that help in maintaining the quality of agricultural produce and greatly reduce food, water and energy waste during its sojourn from farm to kitchen. Apeel is edible with no untoward side-effects. It shall help to improve quality, reduce agricultural waste and ensure overall sustainable future for our planet.

### b. Cold Sterilization

Cold sterilization of agricultural produce and milk has been the mainstay of efforts from the Government of India. Maintenance of cold chain is important to prevent the post-harvest losses considerably. However, in milk, there is a natural system, *i.e.*, lactoperoxidase and thiocyanate, that if exploited can help reduce the post-harvest losses significantly and needs no major infrastructure.

### c. AI for Preventing Post-harvest Losses

Artificial intelligence (AI) can also be helpful for controlling post-harvest losses and meticulous marketing of agricultural produce. AI, IoT and ANN can be integrated into a perfect amalgam for efficient marketing of agricultural produce.

Precise prediction of shelf life based on moisture and other attributes of products can help reduce the post-harvest losses. After calculating probable shelf life, the produce can be marketed at nearby or far off places with minimum chances of wastage.

Image recognition and thermal imaging are other important interventions that aim to analyze the quality of agricultural produce using vision technology. Based on results, further processing may be planned accordingly for rich benefits. However, these technological interventions shall need efficient implementation at each of the producer, processing, logistics, retailer, store and consumer levels. Block-chain technology shall help at improving the transparency in the agricultural sector, the most critical challenge.

## 6. Next Generation Biotechnology-Genomics

Next generation biotechnology and genomics with intense application have tremendous scope in Indian agriculture, particularly the livestock sector. Big data analysis aimed at better understanding of production and reproduction traits and diseases with increased stress on functional genomics shall help transform the agrarian economy at the earliest possible time. There is an urgent need to integrate different *omics*-branches *i.e.*, genomics, transcriptomics, metabolomics, phenomics and ionomics. At most of the instances, only a single or few *omics* branches are stressed in research and development (R&D) initiatives in India that limit overall benefits. Establish gene banks and data banks are other important scientific interventions aimed at securing immense biodiversity in India. It shall also help in reaping rich benefits, in the long run, keeping in view the looming threat of global warming in the present era. There is an urgent need to prioritize biodiversity and adaptation traits in R&D initiatives in Indian agriculture. In conventional breeding programmes, the adaptation traits should be given due consideration.

## 7. Use of Bio-peptides and Water Retaining Superabsorbent Polymers

Bio-peptides or bioactive peptides refer to specific protein fragments with a positive impact on body functions or conditions and may ultimately influence health (Kitts and Weiler, 2003). Plant bio-peptides are useful for a variety of functions in plant, animal and human lives. These uses include enhancing host defense against pathogen infection, regulation of growth and development and human health maintenance.

Bio-peptides promise huge potential for improvement of agriculture produce and animal welfare by helping in disease diagnosis and therapeutics. Bio-peptides are particularly helpful for diagnosis and treatment of mastitis in animals. Besides, bio-peptides have been established to possess anti-proliferative, anti-microbial properties, blood pressure-lowering effects, cholesterol-lowering ability, anti-thrombotic and anti-oxidant activities, enhancement of mineral absorption/ bioavailability and opioid-like activities (Zambrowicz *et al.*, 2013). Bio-peptides have a big role to counter the threat of anti-microbial resistance developing in animals as well as humans (Park *et al.*, 2011).

Soil conditioners called super absorbent polymers can prove to be of immense help in producing agricultural produce in water-stressed areas. They are replinishable and can be used again after recycling. They provide precise amount of water to the plant without wasting much of it. In India, there are water-stressed areas where these polymers will be of golden value.

## 8. Innovative Greenhouses

Greenhouses have been an integral part of the agricultural system in India. Innovations can be directed towards improvement of greenhouse architecture so that produce from the same greenhouse is increased. These include multi-deck greenhouse farming and using aeroponics and hydroponics in greenhouses in a multi-tier fashion. Self-heating greenhouse can be an important innovative intervention that is equipped with ground heat air transfer system aim to use thermal energy of the ground. During summers, the overproduced heat in the greenhouse may be diverted through specially designed pipes to the bottomland. During winter and cooler nights, the exhaust fans work in the opposite direction so that thermal energy is re-used for heating the greenhouse. This will also facilitate the off-season cultivation of fruits and vegetables in these greenhouses.

## 9. Innovative Use of Smartphone Technology

Phenomics and its recording are not at its best in India. This is one of the major factors that limit the application of scientific breeding and management in Indian agriculture. This limitation negates the efforts of progeny testing, genome-wide association study (GWAS), genomic selection (GS) and other applied fields. Several Android software applications and kiosks have been developed in India aiming at efficient knowledge dissemination to farmers. However, enough R&D efforts have not been directed towards facilitating correct and reliable phenome recording in unorganized farms. Therefore, to boost easy phenomic recording, an innovative intervention using chatbots, Google assistant type algorithms and apps in line with Google voice can be used. By this way, the information recording from the small and unorganized farmers can be made simple. The problem of smallholding in Indian agriculture may need encouragement through government policies by providing help in terms of subsidies and mechanization only to self-help groups.

## 10. Other Innovative Interventions

Development of farm management and training softwares is yet another intervention that may prove helpful wherein the farmers will be trained directly. In relation to this, Plantix App has been recently developed for disease diagnosis in plants. These softwares and Apps shall aim at recording and analysis of farm parameters in real-time so that early and pro-active steps may be taken for preventing any untoward happenings.

Use of solar power in agriculture is the bare minimum and no major efforts have been made regarding the intense applications of solar powers for agricultural mechanization. Solar power may be a very useful intervention for desalination of sea water; which may be subsequently used for irrigation. It shall improve agriculture and help in the reduction of carbon footprints.

The approach of thermal imaging and thermography has wide applications in agriculture and allied sectors. It is aimed at efficient assessment of water content of agriculture land, better soil health and management, nursery and greenhouse monitoring and informed predictions *vis-à-vis* disease incidence.

The approach of zero liquid discharge is another important intervention that uses the phenomena of ultrafiltration and reverse osmosis. Particularly useful in the fisheries sector, this approach may help counter water shortage problem in fish rearing. Biological pest control can be applied in which predatory mites are used as biological agents for controlling pests and weeds. For instance, the Ladybird beetles may act for biological control of pests in greenhouses. They leave no residues in the environment.

## Conclusion

Indian agriculture is a major enterprise supporting a huge population base. The transformation of agrarian economy will need immense efforts and innovative interventions from all the stakeholders *i.e.,* producers to consumers. The author has tried to put his viewpoints which may have a huge impact on the agricultural economy in our scenario and help achieving the target of zero hunger at the earliest possible time. However, the only point that the author would like to stress is that, for transforming the agrarian economy, we shall need to think like a Queen or a King of agriculture but work like a slave so that dream of sustainable and ever-prosperous agriculture is realized. If there is any best time to do so, it's now, it's now and it's now.

# Chapter 21

# Strategies to Achieve Zero Hunger through Agricultural Technologies

**Vijaymahantesh and Vijaya R. Chitnis**

*Directorate of Extension,*
*University of Horticultural Sciences, Karnataka*

'**Hunger**' the word itself brings the picture of the strong urge to eat into our minds. Practically causes of hunger are not underlying diseases but, lack of food and nutritional availability. By definition hunger can be described as "a craving or urgent need for food or a specific nutrient" (https://www.merriam-webster.com/dictionary/hunger). Hunger can also be further divided in to acute, chronic, and hidden hunger based on its prevalence. The recent data from food and agriculture organization suggests, that more than 690 million people are suffering from one or the other farms of hunger, which accounts about 8.9 per cent of the world's population (http://www.fao.org/publications/sofi/2020/en/). The beginning of 21st century have found steady decline in this percentage, however from 2015 onwards the numbers are spiking up again indicating the serious concerns. The facts and figures of United Nations suggests "Nearly 10 million people in one year and by nearly 60 million in five years suffer from hunger. Among the majority of the world's undernourished (381 million) are still found in Asia, whereas more than 250 million suffering from hunger live in Africa, where the number of undernourished is growing faster than anywhere in the world. Estimates also suggest that in 2019, close to 750 million-or nearly one in ten people in the world were exposed to severe levels of food insecurity. An estimated 2 billion people in the world did not have regular access to safe, nutritious, and sufficient food in 2019. If recent trends continue, the number of people affected by hunger will surpass 840 million by 2030, or 9.8

per cent of the global population. 144 million children under age 5 were affected by stunting in 2019, with three quarters living in Southern Asia and sub-Saharan Africa. In 2019, 6.9 per cent (or 47 million) children under 5 were affected by wasting, or acute undernutrition, a condition caused by limited nutrient intake and infection (https://www.un.org/sustainabledevelopment/hunger/).

## Agriculture or the Nutritious and/or Balanced Food can Play Paramount Role in Solving Problems Related to Hunger

Some of the measures or goals to achieve zero hunger through agricultural technologies may include, doubling the agricultural productivity and the farm income by 2030. This can be achieved by supporting the small and medium farmers or agro-producers by various means such as integrating dairy, goatery and fishery for sustainability. Additionally, provide non-farm empowerments like financial services, knowledge of inputs, and value addition to agricultural products for productivity and better nutritional status (https://www.un.org/sustainabledevelopment/hunger/.)

Crop productivity is greatly affected by climate abnormalities across the world (Maiti and Satya, 2014; Pandey *et al.*, 2017). The climate change affects rainfall patterns, precipitation amount causing frequent drought spells (Lobell *et al.*, 2011). Certainly, research activities have been directed towards breeding of resistant cultivars globally. However, the minimal success has been achieved with the molecular breeding approach such as fine-tuning of critical genes. Additionally, cutting-edge new technologies such as transgenic approach and omics-based approaches are relatively complex and need specialized skill-sets and a big budget to carry out the experiments.

Keeping above things in mind it is paramount to develop simple, non-expensive yet effective techniques for developing crop plants with enhanced stress tolerance. To maintain the ecosystem and soil health by boosting the tolerance to drought, flooding, and biotic stresses.

Given that, we have lot to achieve through cutting-edge new technologies, however we also should pay equal attention to have genetic diversity of crop plants and farm animals. It is necessary in changing climate and weather patterns to develop crops which have not only improved agronomic traits but also having better nutritional value. Having diversity with respect to crops plants and animals would help in breeding to address new challenges. Additionally, having gene bank, seed banks at various levels with ease of access to the scientific community would further strengthen research activities in this area.

Popularizing scientifically proven Indigenous traditional knowledge for various agro practices like seed storage and medicinal plants would empower local community as well.

Choosing the alternative crops which provide better diet value such as minor millets over rice and wheat. Despite increase in the popularity of minor millets in recent past, the expected reduction in the production of staples is not incredibly significant. Since minor millets have better nutritional value compare to major food grains one must promote the production. Some of the major points includes, developing high yielding cultivars, scientific agricultural practices, post-harvest, and processing technological advancements *etc.*

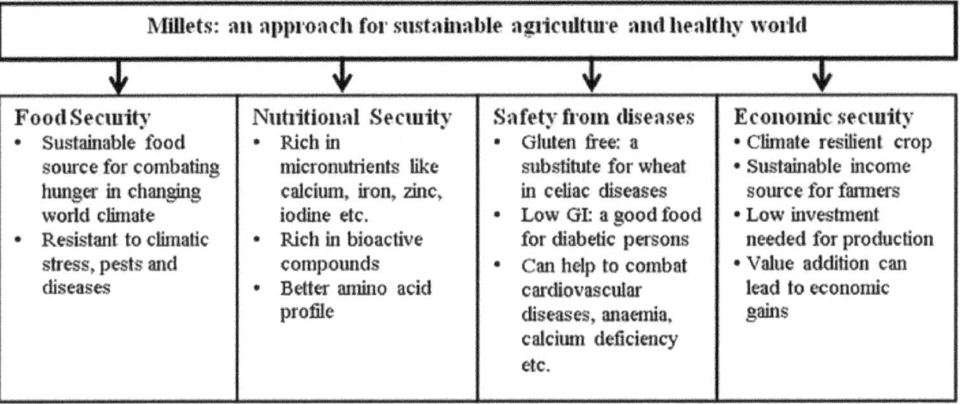


**Figure 21.1**

*Source*: Kumar, A., Tomer, V., Kaur, A. et al., Millets: a solution to agrarian and nutritional challenges. *Agric and Food Secur* 7, 31 (2018). https://doi.org/10.1186/s40066-018-0183-3.

Another vital aspect to sustainable production is maintaining soil health and plant nutrition, in the coming decades soil health going to play major role in production of food crops. Given that more and more arable land is getting spoiled due to bad management practices, one should address this issue first. In this direction having equilibrium in soil ecosystem, which harbors beneficial organisms is paramount. Taking into the account of underdeveloped and developing countries, one must also investigate simple and non-expensive techniques in developing crop plants with increased stress tolerance. In this direction the "HOLOBIONT" concept (Reviewed by Rosenberg and Zilber-Rosenberg, 2018) resonate very promisingly where the beneficial effect of microorganisms can be exploited to enrich host system with it, to induce increased tolerance against biotic and abiotic factors.

# REFERENCES


FAO. UNICEF. WFP. WHO. 2017. The State of Food Security and Nutrition in the World: Building Resilience for Peace and Food Security. Food and Agriculture Organization of the United Nations (FAO); Rome, Italy: 2018. http: //www. fao.org/publications/sofi/2020/en/

Kumar, A., Tomer, V., Kaur, A. *et al.*, 2018. Millets: a solution to agrarian and nutritional challenges. Agric and Food Secur 7, 3.

RK Maiti and Pratik Satya. 2014. Research advances in major cereal crops for adaptation to abiotic stresses, GM Crops and Food, 5: 4, 259-279, DOI: 10.4161/21645698.2014.947861.

Rosenzweig C., Elliott J., Deryng D., Ruane A.C., Müller C., Arneth A., Boote K.J., Folberth C., Glotter M., Khabarov N. 2014. Assessing agricultural risks of climate change in the 21st century in a global gridded crop model intercomparison. Proc. Natl. Acad. Sci. USA. ;111: 3268–3273. doi: 10.1073/pnas.1222463110.

# Chapter 22

# Safety and Quality Assurance of Animal Food Products: Key to Achieving Zero Hunger

**Atul Kumar**

*DGCN College of Veterinary and Animal Sciences,*
*CSK HP Agricultural University, Palampur – 176062, Himachal Pradesh*

## Introduction

Unsafe food is a threat to human health and economies globally. Each year worldwide, unsafe food causes 600 million cases of food borne illnesses and 420 000 deaths. About 30 per cent of food borne deaths occur among children under 5 years of age. Currently, 161 million children under the age of five years old also have stunted growth, while a lack of essential nutrients can damage the health and futures of generations of youngsters. Malnutrition can impact their learning capacities, thereby potentially restricting what their futures look like. This means access to foods that are nutrient dense, such as animal proteins, is key. World Health Organization (WHO) estimated that 33 million years of healthy lives are lost due to eating unsafe food globally each year, and this number is likely an under estimation. Therefore, ensuring food safety is a public health priority and an essential step to achieving food security. Effective food safety and quality control systems are key not only to safeguarding the health and well-being of people, but also to fostering economic development and improving livelihoods by promoting access to domestic, regional and international markets. Safe food production enables market access and productivity, which drives economic development and poverty alleviation, especially in rural areas. Collaboration across sectors and borders is essential to keeping food safe along the entire supply chain. Therefore, food safety is an important key to achieving 'Zero Hunger'.

Although food availability has increased along with the growing human population over the last 30 years, there are still 800 million people suffering from malnutrition. This problem is not only the result of insufficient food production and inadequate distribution, but also of the financial inability of the poor to purchase food of reasonable quality in adequate quantities to satisfy their needs.

With more than 50 per cent of the world's extreme poor living in rural areas dependent on agriculture, maintaining healthy livestock offers a path out of poverty, starvation and malnutrition. Livestock production constitutes a very important component of the agricultural economy of developing countries, a contribution that goes beyond direct food production to include multipurpose uses. Furthermore, livestock are closely linked to the social and cultural lives of several million resource-poor farmers for whom animal ownership ensures varying degrees of sustainable farming and economic stability.

Livestock are multifunctional in developing countries where it plays a pivotal role in farming systems by providing food and income, draught power, fertilizer and soil conditioner, household energy and a means of disposing of otherwise unwanted crop residues. Globally, livestock sector is highly dynamic and is a major industry with more than 12 per cent of the world's population depending solely on it for their livelihood [1].

India's livestock sector is one of the largest in the world. Livestock have been an integral component of our agricultural and rural economy since time immemorial both at the local as well as at regional and national levels. It is now more valued as source of food and contribute over one-fourth to the agricultural gross domestic product and engage about 9 per cent of the agricultural labour force. In recent years, the livestock sector has been growing faster than crop sector which is evidenced by the fact that in 2010-11, livestock generated outputs worth Rs 2075 billion (at 2004-05 prices) which comprised 4 per cent of the total GDP and 26 per cent of the agricultural GDP. The total output worth was higher than the value of food grains [2].

Among various contributions of livestock, foods of animal origin such milk, meat, eggs and fish have always been a constituent of human diets. Nutritionally they are important sources of protein of good quality and excellent sources of vitamins and minerals. However, livestock disease threatens these vital resources, limiting the availability of safe, nutritious food. Nowadays, the factors such as increased consumer interest in high protein diets and increasing awareness and availability of value added retail products through organized food chains are driving the demand for animal origin foods. However, food is also the major source of human exposure to various contaminants. Milk, meat and eggs are rich in protein and essential components of a nutritious diet.

Therefore, concerns regarding food safety especially for foods of animal origin (milk, meat, eggs, fish as well as honey) are increasing worldwide [3]. Thus, current issues pertaining to food safety and quality assurance of animal products has direct proportional linkage with the market value of animal products being produced/supplied by landless, small and marginal farmers especially in developing nations like India.

## Current Status and Potential of Indian Livestock Sector

The livestock sector has emerged as a vital sector for ensuring a more inclusive and sustainable agriculture system. Evidence from the National Sample Survey Office's 70th round survey showed that more than one-fifth (23 per cent) of agricultural households with very small parcels of land (less than 0.01 hectare) reported livestock as their principal source of income. Farming households with some cattle head are better able to withstand distress due to extreme weather conditions.

### Dairy Industry

India is the oyster of global dairy industry that provide opportunities galore for the entrepreneurs globally. Since, last 15 years, India continues to be the largest producer of milk in the world. The milk production in India has significantly increased from 97.1 million tonnes in 2005-06 to 187.7 million tonnes in 2018-19. Similarly, the per capita availability of milk also increased from 241 grams in 2005-06 to 394 grams in 2018-19. This represents sustained growth in the availability of milk and milk products for our growing population. About 70 million rural households are engaged in dairying in India. The consumption of milk is rising, commensurate with an increase in the purchasing power of people, increasing urbanization, changing food habits and life styles and demographic growth. Moreover, milk is the only source of animal protein for the largely vegetarian population of the country.

### Meat and Egg Industry

Among animal foods, meat is considered as highly nutritious and has become an integral component of human diet being a rich source of valuable proteins, vitamins, minerals, micronutrients and fats. Meat consumption is supposed to supply omega 3 fatty acid and conjugated linoleic acid that affords multifaceted nutrient for human health. India stands 5th in the meat production with production of about 6.3 million tonnes. The contribution from poultry, sheep and goats accounts for 31 per cent of total meat produced in India. The majority of meats consumed in India are poultry followed by mutton, chevon, pork and fish and they have emerged as a good substitute for beef.

Poultry farming is by far, most profitable business especially when it comes to feed conversion efficiency and net profits. Poultry meat and eggs are rapidly becoming the major source of animal protein in the diets of Indian consumers which is evident from the fact that per capita consumption of poultry meat in India during 2011-12 was estimated to be just 1.3 kg/person/year but when this is multiplied by the human population of some 1.2 billion, the total quantity of poultry eaten was massive *i.e.* 2.2 million tonnes. Currently poultry meat uptake is considered to be about 3 kg/person, and the poultry industry considers that this will get tripled to 9 kg by the year 2030. To meet this ever increasing demand of inexpensive and safe eggs and meat, the Indian poultry industry has also shown tremendous growth during the recent past.

Therefore, all in all, livestock and their products in the form of milk, meat and eggs contribute significantly towards livelihood and income of farmers in India.

## Food Safety and Quality Assurance

Absolute safety is an unattainable goal for any food. However, food is considered to be safe if there is reasonable demonstrated certainty that no harm will result from its consumption under anticipated conditions of use [4]. This is the biggest driving factor that determine the demand and price of food commodities in current scenario. Nowadays, public is paying much attention to the issues of food safety and food quality and thus demand for foods having these concerns in mind has consistently been increased. This increase in demand for high quality foods, including food of animal origin is strengthened by an increase in overall household annual income at each level of income-brackets.

Similar to any type of food industry, food safety and food quality are two important aspects in the livestock industry. Food safety and quality are the totality of characteristics of the food products that bear on their ability to satisfy all legal, customer and consumer requirements [5]. These are two different concepts that collectively have the ability to ensure consumers that the product has met the industry's standards to be purchased. It is noteworthy that food safety is not synonymous with food quality, although there might be an overlap. Food quality includes all product attributes that influence its value to consumers (*e.g.* freshness, colour, flavour, texture, taste, nutritional value, the animal welfare, fat content of foods, environmentally friendly production, and sustainable farming practices *etc.*) [6, 7]. Whereas, food safety includes all measures intended to protect human health *i.e.* pathogenic microorganisms, misuse of food additives and contaminants such as chemical or biological toxins and adulteration are prevented in food. Although these two concepts have fundamentally different aspects, their collegial meanings make consumers purchase certified foods defined as one that has passed through various stages of the integration traceability methods and quality control systems [8].

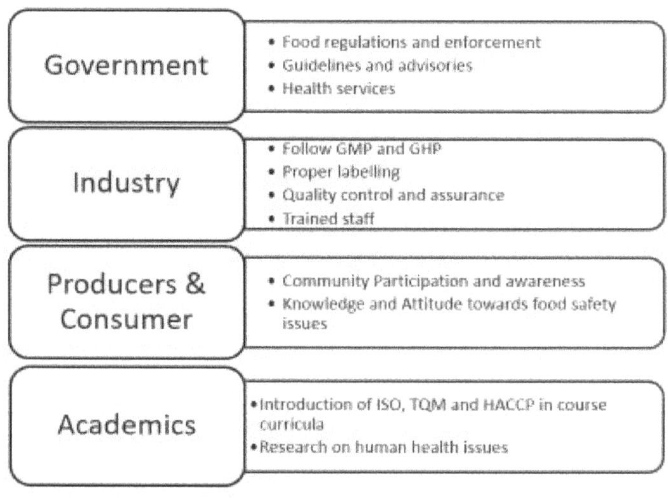


**Stakeholders of food safety**

The food industry uses various systems for quality and safety management *e.g.* ISO 9000, Total Quality Management (TQM), Hazard Analysis and Critical Control Point (HACCP) *etc.* These systems are very effective, but complex and expensive. However, there are certain techniques which are simple and low-cost for food quality control. In India, there are well defined food laws that regulate the players in the food chain from "farm to fork". The national food safety and quality system is regulated by Food Safety and Standards Authority of India (FSSAI). Food Safety and Standards Act (FSSA) 2006 has been implemented from 5th August 2011. With the coming into effect of FSSA, the Prevention of Food Adulteration Act, 1954, and all acts and rules on milk and meat adulteration and quality stand repealed. As per FSSA 2006, no article of food shall contain pesticides, veterinary drugs and antibiotic residues and microbiological counts in excess of such tolerance limits as may be specified by regulations [9]. It aims at promoting public health, and protecting the consumers against health hazards, and enhancing economic development. Despite this, testing of milk for safety and quality parameters at the collection centres is almost non-existent. The adulteration of inferior with superior quality meat is also a common practice for financial gains in India. This has tarnished the India's global image in the export market directly affecting the income of associated stakeholders including landless, small, marginal, and progressive farmers as well.

## Role of Food Safety and Quality Assurance

Effective national food control systems are essential to protect the health and safety of domestic consumers and for overall economic development of the nation and its farmers. National food control systems are also critical in enabling countries to assure the safety and quality of their foods entering international trade and to ensure that imported foods conform to national requirements. The new global environment for food trade places considerable obligations on both importing and exporting countries to strengthen their food control systems and to implement and enforce risk-based food control strategies. Consumers are taking unprecedented interest in the way food is produced, processed and marketed, and are increasingly calling for their Governments to accept greater responsibility for food safety and consumer protection. On quality awareness, offering fresh and cheap produce for sale is no more enough. One cannot just buy milk from a farm and export. Food safety and quality regulations of most countries need proof of traceability. This means keeping records that prescribed norms have been followed at farm, storage, packaging and transportation levels. Therefore, a food safety and quality assurance system plays various roles at different levels.

## Economics Development in Domestic Market in Relation to Food Safety and Quality Assurance

There is a huge demand for safe, high-quality foods with a long shelf-life. However, food of animal origins are biochemically unstable, *i.e.* they deteriorate very quickly. Good quality products cannot and can never be made from poor quality raw milk, meat and eggs. Therefore, good quality raw material must be: free from debris, sediments, off-flavors, antibiotics and chemical residues, low in bacterial numbers, and normal composition. Since, there have been reports on occurrence of contaminants in food of animal origin, its market demand may drastically change as a result of consumers mistrust in both the quality and safety of the product. Therefore, food testing and quality control is an essential component of any food processing industry whether small, medium or large scale. A quality control system encompasses both food safety and quality assurance. It tests food products for safety and quality, and ensure that collectors, processors and marketing agencies adhere to accepted codes of practices and follow the correct methods. The chain of event and advantages that can occur due to implementation of proper food safety and quality assurance system could be as follows:

1. With a good quality control system, farmers can get a fair price in accordance with the quality of their produce thereby enhancing his income per unit of food produced.

2. The processor who pays the farmer can be sure that the raw food is of good quality and is suitable for making various products.

3. Consumers will pay a fair price, *e.g.* moderate price for medium quality, high price for excellent quality.

4. With a good quality control system, the government can protect the health of consumers, prevent contaminated and sub-standard products, and ensure that everyone pays or receives a fair price.

All this is possible only if we have a proper system for quality testing and assurance, which conforms to national or internationally acceptable standards.

## India's Export Potential and its Relation to Food Safety and Quality Assurance

Global market for animal products is expanding fast, and is an opportunity for India to improve its participation in global market. Exports of animal products represent an important and significant contribution to the Indian agriculture sector. The export of animal products includes buffalo meat, sheep/goat meat, poultry products, animal Casings, milk and milk products and honey *etc.* [10]. India's exports of animal products was Rs. 29,813.69 Crores in 2017-18, which include the major products like buffalo meat (Rs. 26033.83 Crores), sheep/goat meat (Rs. 835.75 Crores), poultry products (Rs. 552.16 Crores), dairy products (Rs. 1196.19 Crores), animal casing (Rs. 327.44 Crores), processed meat (Rs. 9.91 Crores), albumin (Rs. 83.72 Crores) *etc.* [11]. There is also huge demand of Indian ethnic meat products in the international market. Further, India is surrounded by the countries which are deficit in production of livestock commodities to meet their domestic demand and thus, India has the opportunity to export livestock products to these countries as well.

Along with several other factors, safety and quality of foods of animal origin are a priority for the industry because higher standards of food safety assurance are being required by our society. But, lack of proper bio-security measures, adulteration of food items, prevalence of infectious diseases and subsequently indiscriminate drug usage and failure to strictly implement food safety and quality assurance system has not only made these foods unsafe for human health but this has also led to considerable deceleration of economic growth. Further, India may be constrained in having an easy access to developed country markets due to stringent food safety and quality standards followed there. This has an overall negative impact on India's export potential. In the past, the presence of various contaminants in milk, meat, eggs, honey *etc.* in even minute quantities has created problems to the export industry.

There are several examples where Indian livestock industry has suffered huge economic losses due to failure of food safety and quality assurance system in place. Some of them are:

☆ Despite the fact that India is leading meat producer and exporter, its share in the international poultry, chevon and mutton meat trade is very meagre. One of the major reasons for this dismal scenario is the poor product quality of the meat produced particularly in relation to chemicals residues present in it. Although, European Union (EU) lifted a five-year ban on the import of poultry meat from India, but given the quality specifications are very stringent, very few exporters can qualify for the export norms set up by the EU.

☆ The volume of dairy exports to the world trade (2008-09) was approx. 70,800 tons only. Other than ghee, several traditional Indian dairy products seem to have a considerable potential for export which can be realized only through the mechanized and hygienic production, as well as by the application of latest packaging techniques to ensure the quality and safety of the products. India can export more dairy products by improving quality standards and creating an integrated supply chain.

Therefore, food safety and quality assurance system are now becoming integral part of our domestic and export markets. Thus, affecting the overall economy of farmer's engaged in this sector. Hence, if such systems are followed in letter and spirit then this sector can almost double the farmer's income through exploiting full export potential of Indian livestock sector.

## Lack of Food Safety/Quality Assurance and Losses Due to Adulteration and Food Borne Illnesses

Consumers, particularly in developing countries such as India, are often exposed to wilful adulteration of their food supply. This can lead to health hazards and to financial losses for the consumer which could be traced back to the point of origin *i.e.* at the farm level. Adulteration of milk and milk products, meat and meat products, eggs, honey *etc.*, and the use of colors to mask product quality to cheat the consumer are quite common. Although risks associated with adulteration are usually low, such episodes invoke public outrage and anger as it violates public trust in the integrity of the food supply. With 60-70 per cent of the income of middle class families in developing countries being spent on food, food adulteration can impact heavily on both the family budget and the health status of the family members [12]. Therefore, enforcement of food safety laws is must to avoid the economic losses to farmers as a consequence of reduction in the monetary values of their produce.

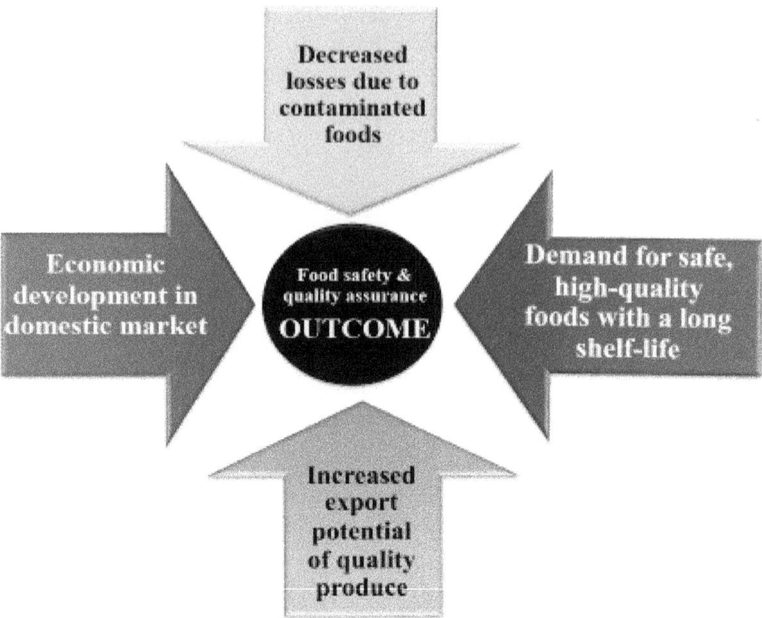


Further, food borne diseases due to microbial pathogens, biotoxins, and chemical contaminants (antibiotics, pesticides, heavy metals *etc.*) in food represent serious threats to the health of thousands of millions of people. Serious outbreaks of food borne disease have been documented on every continent in the past decades, illustrating both the public health and social significance of these diseases. Consumers everywhere view foodborne disease outbreaks with ever-increasing concern. Outbreaks are likely, however, to be only the most visible aspect of a much broader, more persistent problem. Therefore, foodborne diseases not only significantly affect people's health and well-being, but they also have economic consequences for individuals, families, farming communities, businesses and countries. These diseases impose a substantial burden on healthcare systems and markedly reduce economic productivity. Poor farmers tend to live from day to day, and loss of income due to reduction in prizes of food commodities on account of food borne illness further perpetuates the cycle of poverty. Therefore, strict implementation of food safety and quality assurance can directly affect the economy of farming community.

## Significance of Inedible Animal Products on National Income

Inedible animal products also help in increasing the national income and in the upliftment of the rural farmers. As per one of the recent estimates, India produces about 2.65 crore pieces of buffalo and cattle hides and about 7 crore pieces of goat and sheep skins accounting for about 15 per cent of the world's hides every year. Indian hides and skins are one of the best in the world and are much in demand in the international market. The total leather good exports from India stood at US$ 1.36 billion during 2017-18. Same is the case with the wool, where India stood at

7<sup>th</sup> position in the world in terms of wool production contributing about 1.8 per cent to the world's total wool production. Though the wool industry is relatively a smaller sector as compared to cotton or man-made fibre industries, it still plays an important role in the economic development and textile exports of the country. This has been possible primarily due to proper sheep care and cross breeding of the high yielding fine quality exotic breeds with low yielding coarse indigenous breeds. However, the average yield of wool per sheep is still very low as compared to that obtained in Australia, New Zealand and in some other developed countries *e.g.* A Bikaneri sheep gives an average of 1 kg per clip while a fine Merino sheep gives 3 to 15 kg per clip. In India, all of these industries are primarily rural based, export oriented in which the organized sector, the decentralized sector, and the rural sector complement each other. Despite 3<sup>rd</sup> rank in sheep population, India's domestic produce (43.6 thousand tonnes) is not adequate and the wool industry is still dependent on imported raw material. Therefore, India has to import a large quantity of wool, especially the superior quality wool. Hence, to meet the domestic and international demands, these livestock product based enterprises provide great opportunities and avenues for the farmers. If quality assurance of these inedible products is ensured, then in coming years it will significantly enhance the income and livelihood of farmer's engaged in such animal husbandry practices.

## Conclusion

Global and domestic markets trends hints that when the issues of food quality and safety matter, consumers have no doubt that they will have to pay a premium price to purchase safe and high quality products. Together with this, livestock sector is expected to emerge as an engine of agricultural growth in the coming years and beyond in view of rapid growth in demand for animal food products. Further, being one of the largest producers of most of the livestock products, India has the potential to significantly increase and expand the export of livestock products. To give a boost to livestock exports, compliance with various sanitary and phyto-sanitary measures should be taken up vigorously to ensure international hygiene standards and to harness the untapped potential of exporting to developed countries like EU, USA, Australia and Japan. Furthermore, it seems that domestic policy initiatives, increased production and productivity and ensuring food safety and quality are the important factors in enhancing the export of livestock products. As far as domestic market is concerned, the informal sector in India is the major supplier of food products, yet they operate in disregard of food safety and quality controls. This is because the small and medium-sized enterprises and food vendors, in contrast to the formal sector in the food industry, lack qualified personnel, infrastructure and equipment necessary for hygienic storage and handling of food products during production, distribution and retailing. This not only poses a significant risk to the human health but also affects the quality of produce and ultimately its price which has direct bearing on income of farmers. Therefore, department of animal husbandry, dairying, fisheries, in collaborations with agricultural universities and other agencies must actively educate general public and farmers on food safety and quality controls issues. It is well understood that any safety and quality management has cost implications and therefore, income is a major limiting factor. Hence, for addressing top two sustainable development goals *i.e.* poverty alleviation and zero hunger by 2030 through increased animal production system would eventually enhance the farmer's income by stimulating the purchasing power of domestic consumers, consequently promoting hygiene-based demand instead of price-based demand for animal products. Therefore in conclusion, the ways by which livestock industry can supports Zero Hunger could be by enhancing food and nutrition security, protecting vulnerable undernourished children through consumption of animal foods, tackling poverty by acting as a vital source of income to farmers, and by empowering rural communities to be self-sufficient.

## Acknowledgement

Author express his gratitude to Dr. Z.A. Pampori, Professor and Head, Veterinary Physiology, SKUAST – Kashmir for providing the necessary information related to the article during online training programme on: "Achieving Zero hunger by 2030: the critical role of agriculture and allied sectors".

## REFERENCES


1. Food and Agricultural Organization (FAO). World Food Summit. Food for All. Rome 13-17 November 1996 Available at: http: //www.fao.org/docrep/x0262e/x0262e13.htm

2. Report of the working group on animal husbandry and dairying 12th five year plan (2012-17). Government of India. Available at: http: //planningcommission.gov.in/aboutus/committee/wrkgrp12/agri/AHD_REPORT_Final_rev.pdf

3. Kumar A, Gill JPS, Bedi JS, Manav M, Ansari MJ and Walia GS. 2018. Sensorial and physicochemical analysis of Indian honeys for assessment of quality and floral origins. *Food Research International* 108: 571–583.

4. World Health Organization (WHO). Safety Aspects of Genetically Modified Foods of Plant Origin, A report of joint FAO/WHO experts consultation on foods derived from biotechnology. Geneva. 29 May – 2 June 2000. Available at: http: //www.fao.org/fileadmin/templates/agns/pdf/topics/ec_june2000_en.pdf.

5. Will M and Guenther D. 2007. Food Quality and Safety Standards as required by EU Law and the Private Industry with special Reference to MEDA Countries' Exports of Fresh and Processed Fruits and Vegetables, Herbs and Spices. A Practitioners' Reference Book, 2nd Edition. GTZ – Division 45.

6. Lasztity R, Petro-Turza M and T Foldes. 2004. History of food quality standards. In: *Food Quality Standards* (Lasztity R Ed.), In Encyclopedia of Life Support Systems (EOLSS), Developed under the Auspices of the UNESCO. EOLSS Publishers. Oxford, UK. 2004. Available at: http: //www.eolss.net.

7. Nelson MB. 2005. International rules, food safety and the poor developing country livestock producer. Pro-poor livestock policy initiative working paper No. 5. FAO, Rome.

8.  Haghiri, M. 2016. Consumer choice between food safety and food quality: the case of farm-raised atlantic salmon. *Foods* 5(2): 22.

9.  Food Safety and Standards Act, 2006. Available at: https: //www.fssai.gov.in/home/fss-legislation/food-safety-and-standards-act.html

10. Kumar A, Gill JPS, Bedi JS, Chhuneja PK, Kumar, A. 2019. Determination of antibiotic residues in Indian honeys and assessment of potential risks to consumers. Journal of Apicultural Research. 59(1): 25-34.

11. Kumar A. 2010. Exports of livestock products from India: performance, competitiveness and determinants. *Agricultural Economics Research Review*. 23: 57-67.

12. FAO. 2003. Assuring food safety and quality: guidelines for strengthening national food control systems. FAO food and nutrition paper 76. Available at: http: //www.fao.org/docrep/006/y8705e/y8705e09.htm

# Chapter 23

# Battle of Hunger in India: Efforts, Limitations, Possessions and the Way-Forward

Zahoor A. Pampori

*Faculty of Veterinary Sciences and AH,
SKUAST-Kashmir, Srinagar – 190006, J&K*

## ABSTRACT

India became independent in 1947, when it was still reeling from the impact of the 1943 Bengal famine and world as a whole was experiencing the brunt of world war second, thus India was born hungry in a hungry world. The country leaders were well aware of the challenge that India was expected to face in terms of food security and it was Jawaharlal Nehru who said everything can wait but not agriculture. The first president of India Rajendra Prasad after taking the chair, the first thing he did was to raise the flag at the Indian Council of Agricultural Research, declaring "India's most pressing task would be to conquer the battle of hunger. The Indian population has increased tremendously from 376 million in 1950 to 1380 million in 2020 and it is agriculture and its allied sectors that sustained such a huge population. India still has a significant proportion of population 14 per cent undernourished, 35 per cent children stunted, 20 per cent children underweight, 52 per cent women of reproductive age anaemic. India could bring out green revolution, white revolution and blue revolution in order to provide food security to its people. India presently is not food deficient; it has attained self sufficiency in food production and stands exporters of food. However the irony is that India stands at place 102 in global hunger index with score of 30 that is a matter of concern despite India being self-sufficient in food production. The problem is in making this food available to the people. India needs nutritional security rather than food security now, transformation in agriculture and allied sectors to become free from hunger. The task is tough and precipitated by Covid-19 pandemic, but not impossible. India has much strength but will need research, extension, implementation and policy framing to have sustainable, nutrition sensitive, climate resilient, integrated and smart agriculture to eliminate hunger.

## India was Born Hungry in Hungry World

Independent India was born hungry in hungry world on 15th August, 1947. On its birth India was still reeling from the impact of the 1943 Bengal famine that left over 3 million people dead (Maharatna, Arup, 1996). Globally the scenario was

not much different, the world was yet feeling the jerks of world war second that resulted into 39 million deaths in Europe alone. Periods of hunger became more common even in relatively prosperous Western Europe. A severe hunger crisis was witnessed in Greece, Netherlands, Dutch, and Germany during World War II. Impacts of hunger were not restricted to the people living in that era but extended to the individuals exposed to hunger crisis in utero. Thus when India became free from the British occupation in 1947, the whole world was suffering from one or the other form of hunger.

The nature may have shifted a bit with no major famines now but malnourishment remains the real threat today too. India is still struggling to feed its people adequately, according to the UN, India is a home to quarter of the world's undernourished population. Indian leaders and think tanks were aware of the fact that they are going to face tremendous challenge of hunger in coming days after freedom. In India, hunger lay at the roots of much of its public policy in its early years, and over 70 years later, we are still a hungry nation. It was K.A Abbas's 1946 film Dharti Ke Lal, one of the first Indian social-realist films, which spread the message of hunger being a big challenge ahead. The interim government led by Jawaharlal Nehru, which took power in September 1946, realising the hunger challenge, nominated Dr. Rajendra Prasad "President of the Constituent Assembly" and the first thing Dr. Prasad did was to raise the flag at the Indian Council of Agricultural Research, declaring "India's top concern would be to conquer the battle of hunger. Dr. Prasad was made In-charge of food and agriculture on January 26, 1950. Therefore, independent India from the inception of its birth started working towards the food security of its people.

Independence deprived India of wheat and rice as large area of Punjab and Bengal got severed from it. India had to import food as far from Argentina immediately after independence to feed its people whose staple food use to be rice and wheat. Prime Minister Jawaharlal Nehru took pains to tell his fellow Indians that living on rotis made from a mix of wheat and sweet potatoes and reducing consumption of food can be observed as he observed himself in lieu of food insufficiency. Mahatma Gandhi fervently advocated the removal of all controls on food, allowing a total free market to make food available to its people after independence. However, the then Government, reimposed the system of ration cards. In fact it was in British era that World War's forced them to introduce the first structured public distribution of cereals in India through the rationing system-sale of a fixed quantity of ration (rice or wheat) to entitled families (ration card holders) in specified cities/towns. The system was started in 1939 in Bombay and subsequently extended to other cities and towns and by 1946, as many as 771 cities/towns were covered. The Department of Food under the Government of India was created in 1942, which helped in food matters getting the serious attention of the government. When the War ended, India, like many other countries, decided to abolish the rationing system in 1943. However, on attaining Independence, India was forced to reintroduce it in 1950 because of inflationary pressures and high global prices of food grains at the end of the War, which were around four times higher than the pre-war prices (Bhatia, 1985). This intervention with good intentions of reaching to people with food, making food available at a 'fair price' so that access to food gets improved and to keep a check on the speculative tendencies in the market Public Distribution System (PDS) was launched by India. Creation of Food Corporation of India and Agricultural Prices Commission in 1965 consolidated the position of PDS. It made Government to announce a minimum support price for wheat and paddy and procurement of quantities that could not fetch even such minimum prices in the market. The resultant stocks were to be utilized for maintaining distribution through the PDS and a portion of these were used to create and maintain buffer stocks. However, unfortunately bringing in minimum support prices and purchases by government corporations resulted in our current situation where huge amount of grains is languish in government warehouses, while consumers still go wanting. Similarly the land redistribution to the farmers has been part of India's state policy from the very beginning (Thorner, 1976). Independent India's most revolutionary land policy was perhaps the abolition of the Zamindari system. Since independence there has been voluntary and state-initiated land reforms in several states like West Bengal, Kerala and J and K with dual objective of efficient use of land and ensuring social justice (Appu, 1996; Basu and Kaushik, 2008; Besley and Burgess. 2000). Since 1947, India has enacted perhaps more land reform legislation than any other country in the world, it has not succeeded in changing in any essentials the power pattern, the deep economic disparities, nor the traditional hierarchical nature of intergroup relationships which govern the economic life of village society. Indian Government policy was particularly a big failure over land reform that precipitated the fragmentation of land holdings which is being identified as one of the main problems in enhancing agricultural productivity through technology intervention.

## Efforts that India made Post Independence in Food Security

After independence realizing the food scarcity, citizens of India be it industrialists, housewives, farmers or even astrologers kept coming up with ideas

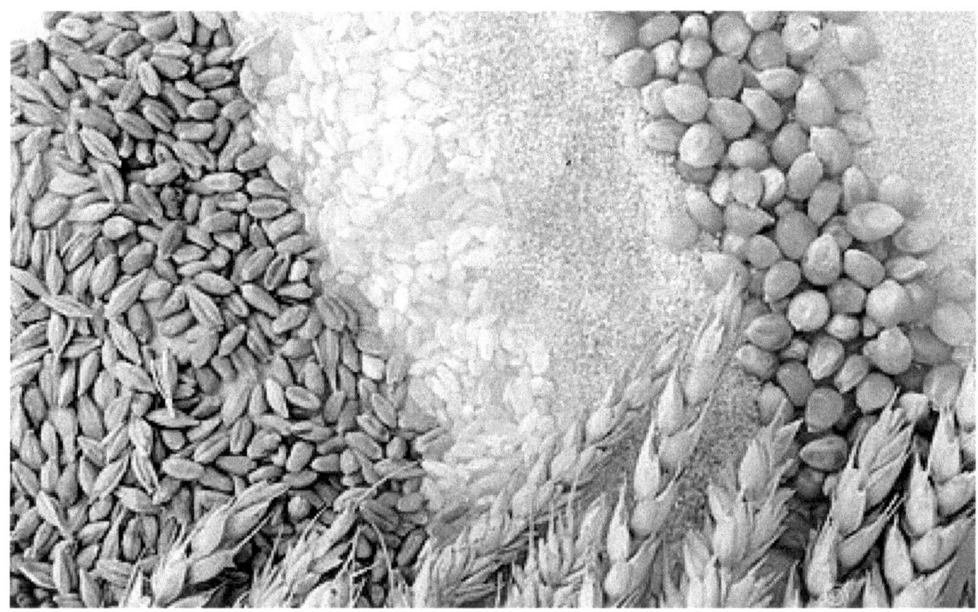

they were sure would solve India's food problem and they were not probably wrong. Monkombu Sambasivan Swaminathan known as father of Green Revolution in 1960s, chooses to study agriculture rather than medicine, rightly judging that abundant food production had an important role to play in keeping a country independent. M. S. Swaminathan undoubtedly helped India to become food self-sufficient from food deficient while launching the green revolution. *The Green Revolution* helped transform the South Asia region from one of food deficits to surpluses and moved millions of people out of poverty (Pingali 2012). Green Revolution was the product of four things,

1. The gene technology that transformed and changed people's understanding of wheat and rice yields,
2. The services that took the technology to the field like extension services, credit and insurance;
3. The public policies of input-output pricing like the prices commission, and
4. The farmers' enthusiasm, that was very important.

In 1999, Swaminathan was one of the only three Indians besides Rabindranath Tagore and Mahatma Gandhi to be on TIME magazine's list of the 20th century's 20 most influential Asians. Green Revolution no doubt increased the yield of staple food per hectare of farm land, three times increase in wheat yield (0.85 tons in 1961 to 2.74 tons 2005) and two times increase in rice yield (1.54 tons in 1961 to 3 tons in 2005) in India (FAO http://economics.cimmyt.org). However, at the same time the population of India recorded tremendous increase from 376 million in 1950 to

1380 million in 2020. This population growth witnessed exponential increase during seventies and eighties with growth rate of 2.5 that has decreased to 0.99 in 2020. To feed such a large population with only little land available for farming, Green Revolution has increased food production from 50.80 million tons in 1950-51 to 176.40 in 1990-91 to 282 million tons in 2018-19 an increase of 455 per cent. India was net importer of food grains and depended upon international food aid upto mid-1960s and is now exporter of food grains to many countries. India is both a large importer and exporter of agri-commodities and has a positive trade balance in this sector. India's exports of agriculture food products increased from US$ 28 billion in 2015 to US$ 31 billion in 2019, and its imports decreased from US$ 20 billion in 2015 to US$ 18 billion in 2019 (World Integrated Trade Statistics).This increase in food production may have made India self-sufficient in food production but not food and nutritionally secure because access to food is still a major issue in India.

Many revolutions successively followed the Green Revolution with objective of achieving food and nutritional security for huge population in the country. These revolutions have achieved the objectives to a larger extent by the technology interventions in agriculture and allied sectors. It was Dr. Verghese Kurien who pioneered the Anand model of dairy cooperatives and replicated it nationwide and brought in **White Revolution** in 1970 and became known as father of White Revolution. White Revolution increased milk production from 20 million tons in 1950-51 to 176 million tons in 2018-19 an increase of 780 per cent. The milk availability in India per capita per day is 394g which is much higher than the ICMR recommendations of 280g. India is world's largest milk producer, it is exporting milk products and has earned 1341 Crores in 2019 (FAO, 2020). However, it is unfortunate that despite being surplus it does not reach adequately to the entire population. Many pockets in India are having availability of milk to over 1000 ml per person per day but many other pockets do not even have three figure availability of milk per person per day.

Yet another revolution referred as *Silver Revolution* started way back in 1970s has increased the egg production tremendously post independence. Production of eggs was only 10 billion in 1950-51 that got increased to 103.3 billion in 2018-19 an increase of 933 per cent. However, owing to the huge country population per-capita availability is still less than half of the recommended *i.e.* 79 eggs/person/year against 180eggs/person/year.

India did not stop at any front to bring it out from hunger and ensure food and nutritional security. In 1980s red revolution was brought in and production of meat got increased from 1.65 million tons in 1961 to 8.10 million tons in 2019 an increase of 390 per cent. India stands third largest exporter of meat in the world today. In 1985-86, **Blue Revolution** in India was one more success story that increased aquaculture production from 44843 tons in 1960 to 8.9 million tons in 2018 an unimaginable increase of 20127 per cent. India stands 2$^{nd}$ in the world for aquaculture production and this sector is fast growing sector with growth rate of over 9 per cent.

India was poor in meeting the oil demands of the country and oil was imported post independence. However, during 1986-87 *Yellow Revolution* was brought in that made India self-sufficient in oil production by 1996-97. However, India has not been able to maintain the production as per the demand and is presently observing a large gap between demand and consumption with per capita availability only 11 g/person/day as against the recommended 30g/person/day. Horticulture sector was targeted in 1990 and India had the period between 1991 to 2003 known as the period of the ***Golden Revolution*** that increased fruit production from 96.50 million tons in 1990-91 to 313.35 million tons in 2019 an increase of 224 per cent (Table 23.1). This made India a world leader in the production of bananas, mangoes, coconut and spices and provided sustainable livelihood and nutrition options. India largest producers of fresh fruits, milk, pulses and oil seeds, sunflower seeds and second largest producers of wheat, rice sugarcane, potato, tea, cotton *etc.* in the world. Today India has self sufficiency in food production and exports various food grains.

**Table 23.1**

| Particulars | 1950-51 | 1960-61 | 1990-91 | 2018-19 |
| --- | --- | --- | --- | --- |
| Population | 380 million | 459 million | 890 | 137 billion |
| Food Grain Production (million tons) | 50.8 | 82.0 | 176.4 | 282 (455 per cent) |
| Milk Production (million tons) | 20 | 20.38 | 54 | 176 |
| Meat Production (million tons) | - | 1.65 | 3.66 | 8.10 |
| Egg Production (Billions) | - | 0.17 | 1.2 | 5.24 |
| Fruit Production (million tons) | - | - | 96.50 | 313.35 |
| Fish Production Aquaculture Capture Fish (million tons) | - | 44843 tons | 1.02 | 8.90 |
| | - | 1.2 | 2.86 | 4.80 |

## Limitations with India to See Hunger Free Country

### Hunger Index

Despite all these leaps, India is still a home to one quarter of world's hungry. 194 million people in India go to bed hungry which accounts for over 14 per cent population of which women constitute 60 per cent. 80 per cent of all hungry live in rural India which ironically stands food producers. India ranks at 102 position in Global Hunger Index (2019) among 117 countries with a score of 30, that puts the India in a group of serious hunger nations. South Asia continues to be placed in the 'serious' category by the Global Hunger Index, with India performing worse than all of its neighbours, barring Afghanistan (Global Hunger Index, 2019).

### Child Mortality

Twenty five million babies are born every year of which 18 per cent (4.54 million) are low birth weight and 0.57 million die every year. Of all these deaths, 88.8 per cent die before first birth day. India has the highest burden of neonatal deaths. Fourty five per cent of under-5 year deaths are due to under-nutrition. India's children are amongst the most malnourished in the world, 20 per cent are wasted, 37 per cent are stunted, 15 per cent under-nourished (Figure). Public Health Foundation of

Global Hunger Index 2019
Rank (Poorest)
GHI Scores
117 Central African Rep. 53.6
116 Yemen 45.9
115 Chad 44.2
114 Madagascar 41.5
113 Zambia 38.1
112 Liberia 34.9
111 Haiti 34.7
110 Timor-Leste 34.5
109 Zimbabwe 34.4
108 Afghanistan 33.8
India is ranked 102 of 117 countries, behind its neighbours Nepal, Pakistan and Bangladesh. 17 countries, including Belarus, Ukraine, Turkey, Cuba and Kuwait shared the top rank with GHI scores of less than 5
India's Rankings
Total Countries
2019 102 117
2018 103 119
2000 83 113
102 India 30.3
94 Pakistan 28.5
88 Bangladesh 25.8
73 Nepal 20.8
66 Sri Lanka 17.1
15.0
Source: Concern Worldwide / Welt Hunger Hilfe

HOW INDIA HAS CHANGED OVER THE YEARS
2000
2005
2010
2019
18.2
22.2
17.5
14.5
17.1
20
16.5
20.8
54.2
47.8
42
37.9
9.2
7.5
5.8
3.9
Undernourished population
Child Wasting
Child Stunting
Child Mortality
Global Hunger Index 2019

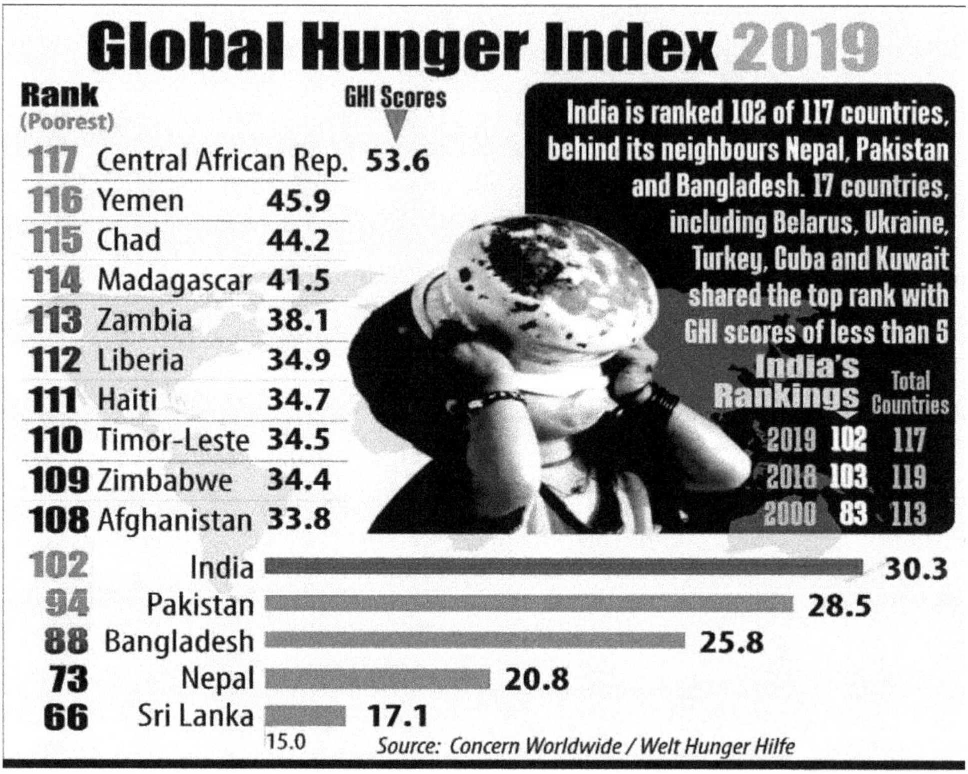

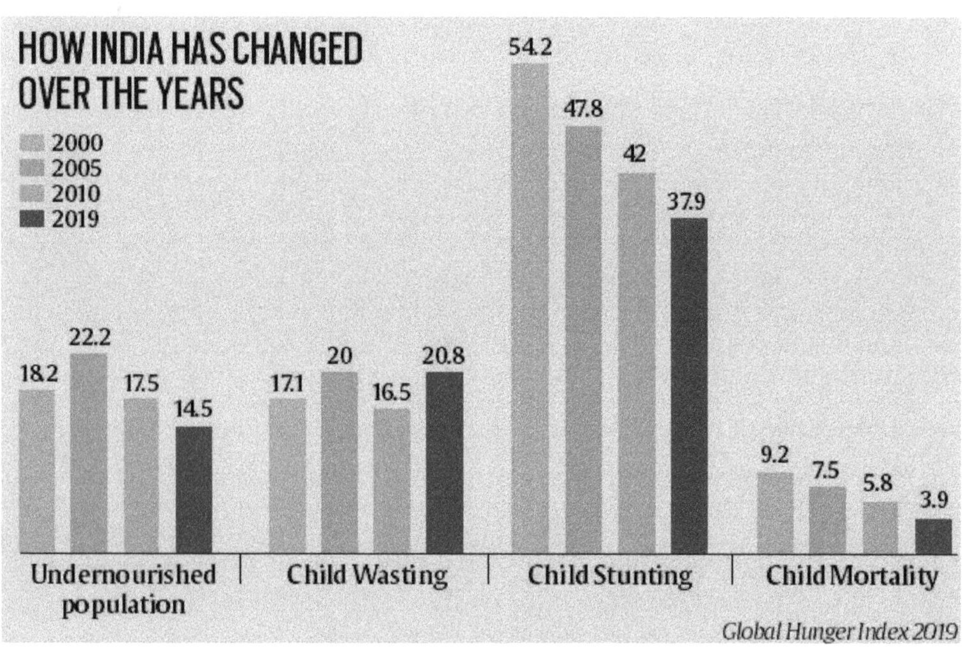

India reported that stunting in children aged under-5 years of age in states ranged from over 50 per cent (Uttar Pradesh) to 19 per cent (Kerala) (Raykar *et al.,* 2015).

## Malnutrition/Under-nourishment

Eighty per cent women of reproductive age are anaemic, 75 per cent women and children are Vit. A deficient. Seventy per cent Indians consume less than 50 per cent RDA of micronutrients and 9 per cent newborn receive adequate diet. The malnutrition or under-nourishment of the children has very serious implications on population as well as national economy. It is the malnutrition that causes neonatal deaths, low birth weight babies, resulting in poor mental health, poor learning and cognitive abilities and poor school performance, increased disease susceptibility, low resistance, high morbidity and mortality, low work productivity, low income and poverty (Singh, 2020; Chatterjee *et al.,* 2016). Women with undernutrition may go through serious pregnancy-related complications (Nguyen, 2019). All these factors ultimately lead to poor work capabilities, low output capacity, low income, poverty and reduced GDP and estimated to cause 11 per cent loss in annual GDP, 22 per cent loss in adult income besides IQ loss of 5-11 points. Poor child nutrition outcomes are generally observed to result from proximal causes such as poor infant and young child feeding practices, poor nutrition among women before and during pregnancy, and poor sanitation practices (Smith and Haddad 2015; Comprehensive National Nutrition Survey, 2019). Therefore, a strong, secure and healthy society contributes to the national economy as well. In addition to factors such as increased globalization and economic growth, the rising trend in purchases of snacks, hydrogenated edible oils and 'other processed foods' have been some of the key drivers of increased consumption of unhealthy processed food products. Further, it has been driven by the growing Indian policy focus on promoting food processing, as well as liberalization of FDI in food processing and retailing, which enhanced the availability and affordability of processed foods (Thow *et al.,* 2016).

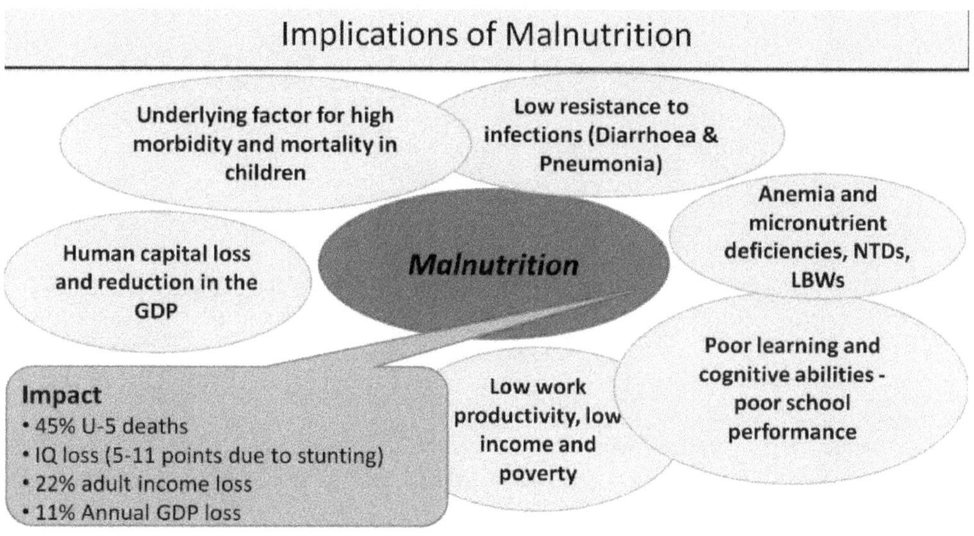

There has been an increased intake of sugar, oils and highly processed food in Indian diets with more apparent changes identified in urban India (Gulati and Misra, 2014). Studies have shown that the new dietary patterns are contributing towards a number of non-communicable diseases like diabetes, hypertension and cardiovascular diseases in India, which account for a major share of total deaths in the country (Prabhakaran *et al.*, 2018). Urbanisation, and changes in people's lifestyles and diets have caused a steady increase in overweight/obesity among people (Dutta *et al.*, 2019; Ford *et al.*, 2017). Current high levels of malnutrition are often due to unbalanced diets with insufficient nutrition diversity that continue to a health challenge in India (Nie *et al.*, 2016).

## Land Status and Food Production

Despite various success stories in increasing the food production parallel to the increase in human population, there are some limitations that come in way of hunger free India. India's agricultural food production has increased from Rs. 9.7 trillion in 2009-10 to Rs. 24.93 trillion in 2017-18, accounting for an overall growth rate of around 28.73 per cent. Within sub-categories, milk products, fruits and vegetables, and cereals (which include rice and wheat) are among the top agricultural produce of India (National Account Statistics, 2019). Among them, output of milk products has grown at a faster rate (66.27 per cent) between 2013-14 and 2017-18 compared to cereals (22.12 per cent). Mahmoud Solh argues that the projected 50 per cent increase in demand for agricultural production from 2013 to 2050 (FAO, 2017) has to be achieved despite the ongoing degradation of natural resources and the serious implications of climate change on agriculture. The very important issue with the country is available land and land holdings. India has only 2.3 per cent of the world's land with 17 per cent of world's human population (2nd largest population in the world). Fortunately 53 per cent of it is arable and 57 per cent is rainfed. However 91 per cent farmers are marginal and small farmers with less than 2 hectare of farm land that infact limits the technological interventions in farming, a major constraint in increasing crop production. Average land holding has reduced

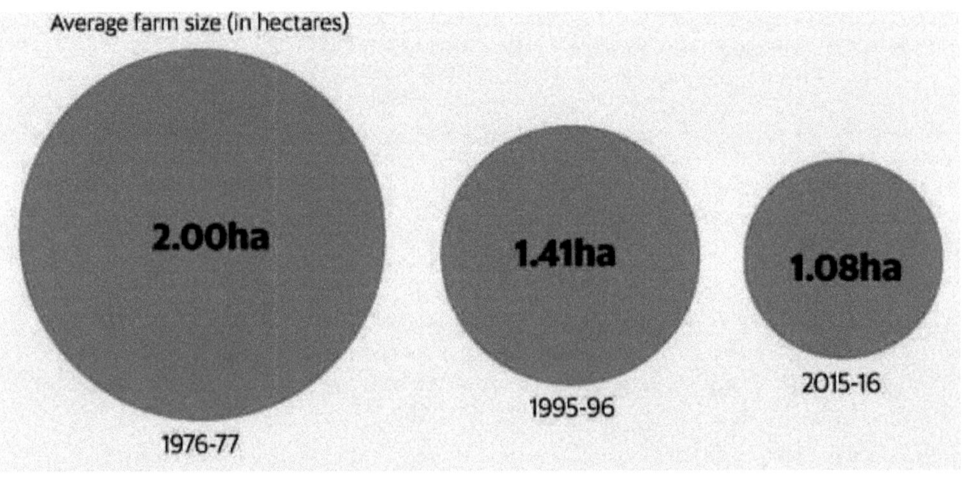

from 2 hectares in 1976-77 to 1.08 hectares in 2015-16. Further the soil has fatigued, organic matter status has declined and micronutrients are deficient. The problems of soil acidity, salinisation and sodification are very common. Soil organic carbon has reduced to 0.3-0.4 per cent, which is four times less than the ideal of 1.5 per cent. India has highest agriculture water withdrawals, 2 times higher than China and 6 times higher than USA and it is projected that India will have 20 per cent water depletion by 2022. Excessive intensification (*i.e.* monoculture) risks simplifying diets and worsening nutrition in producer communities and threatens ecosystem resilience. Expansion of deserts, soil erosion, water scarcity and extreme weather phenomena as a result of climate change are becoming particularly apparent in countries that already suffer from hunger and poverty.

## Farmer's Plight

Besides deteriorating soil conditions for optimal food production, the farmer's plight too precipitates the situation of food insecurity. Ironically, 40 per cent of farmers which are producers themselves are food insecure. People engaged with agriculture farming are more as labourers rather than cultivators, in 1960-61, 76 per cent farmers were cultivators and only 24 per cent labourers but as of now only 45 per cent are cultivators and 55 per cent are labourers. Further 46 per cent of India's total workforce is engaged in agriculture which in turn contributes only 17 per cent to GDP, that in other terms means more input and little output, an inefficient system (Report on Fifth Annual Employment-Unemployment Survey, 2015-16; Press Information Bureau. GOI, 2018). There is an increasing trend in urbanization and India is expected to lead the world in urbanization which means people engaged in agriculture will get reduced and agriculture produce will suffer a lot given the low land holding structure of India. People associated with farming are losing interest in it because there is no proper market link and many a times farmers do not get even the production price to their produce. Lack of storage facilities particularly for perishable produce further aggravate the farmer's plight and farmers sometimes lose everything and we have seen in India farmers committing suicide.

## Food Storage Capacity

India's total storage capacity is 70.5 million tons as against requirement of 80.5 million tones that too is mostly in Utter Pradesh for Potatoes. Available cold storage capacity is less than the half of the required, only 29.7 million tons cold storage capacity as against 61.7 million tons. Transportation of food grains is not efficient and more than 30 per cent of grains supplied through the public distribution system is lost because of inefficient storage and transportation system in India. Studies have shown that farmers have low price realisation and there is a huge wastage in the supply chain due to fragmentation, poor storages, inefficient information flow (Negi and Anand, 2016; Pingali *et al.*, 2019, Gokarn and Kuthambalayan, 2017).

## Livestock Sector

Similarly in livestock sector, India keeps high proportion of low yielding indigenous livestock with no effective policy to address the issue. Artificial insemination for improvement of local germplasm with improved and exotic germplasm has not achieved promising success and India still has only 40 per cent AI success. Embryo transfer technology has yielded scores of research papers in peer journals but has not been impressively implemented in the field with the result there has not been much improvement in the productivity per animal. India has a huge potential of exporting livestock products owing to large livestock population but again due to lack of standard quality control mechanism, inability to adhere to the food safety standards and poor cold chain mechanism, India has not made success in increasing the global market for livestock products that stands only 2 per cent.

## Agriculture Policies

Agriculture system in India is not designed to promote health. There is more focus on increasing productivity and not the nutritive quality that is why India may be food secure but not nutritionally secure. The government both at the Centre and state levels have implemented several programmes and schemes towards 'nutrition security for all', which includes National Food Security Act "Right to Food Act" (2013), the National Nutrition Strategy (2017) and the National Nutrition Mission (2017). The Act made access to food a legal right, targeted Public Distribution System (TPDS), Mid-Day-Meal (MDM) and Integrated Child Development Services (ICDS-Anganwadi-Supplementary Nutrition Programme) but the expected outcome is yet to be relished. The Food Safety and Standards Authority of India (FSSAI) was formed in 2011 to ensure food safety and standards along with food security and nutritional safety. It has taken several initiatives such as the 'Eat Right India' movement on July 10, 2018, which is a multi-sectoral effort to nudge citizens to 'eat right/have the right diet'. Despite the government taking several initiatives to promote nutritional security, studies have pointed out that India suffers from dual problem of malnutrition, *i.e.* both undernutrition and overnutrition (Global Nutrition

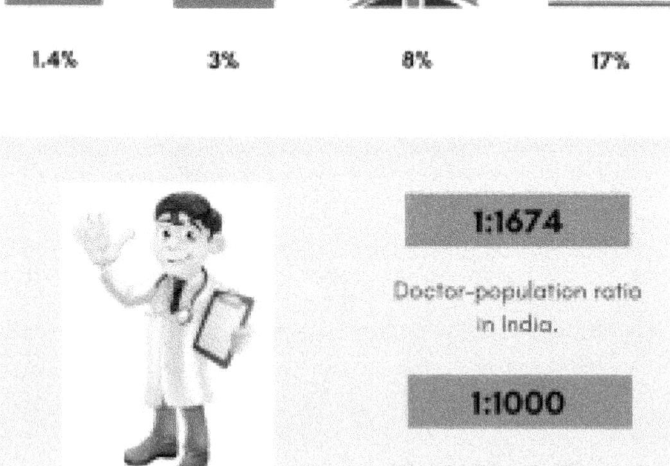

Report, 2020; MOSPI and WFP, 2019; Indian Council for Medical Research and National Institute of Nutrition (NIN), 2019; World Health Organization (WHO), 2018; International Life Science Institute (ILSI) India, 2018). NFHS-5 (2019-20) shows that food security and nutrition in India have worsened since the last NFHS-4 (2015-16). Among the 22 states and Union Territories (UTs) for whom the data was released, 18 show either stagnation or worsening of stunting levels among children less than five years. Since the data presents the pre-Covid-19 picture, the present nutritional status could be more worrying, especially for the poor and the marginalized sections.

Little priority for health sector in India has put it at the bottom of the GHI list. A meagre 1.4 per cent of GDP is spend on healthcare in India less than Nepal (2.3 per cent) and Srilanka (2 per cent), that otherwise stands 17 per cent in America, 8 per cent in Britain and 3 per cent in China with world's largest population (World Bank reports on public expenditure and infant mortality). India is far behind the world average for physicians (0.6 V/S 1.3/1000) and hospital beds (0.9 V/S 3/1000). More than seven in ten Indians are not covered by insurance (National Family Health Survey 2015-16, NFHS-4). Indians are the sixth biggest out-of-pocket health spenders in the low-middle income group of 50 nations (Vipul Vivek, 2017; Swagata, 2017). Compounded by high out-of-pocket expenditures, health-care expenditures exacerbate poverty, with about 39 million additional people falling into poverty every year as a result of such expenditures (Balarajan, *et al.*, 2011). The governments mostly do not align their policies to the needs of the poorest population. There is a lack of strategies to promote agriculture in the country in such a way that no-one goes food and nutritionally insecure. Diversified food basket is missing giving rise to deficiency diseases. Corruption is one of the greatest obstacles to the development and land grabbing is a big problem.

## Gender Inequality

Women form an integral part of the agricultural sector, and in India women make up a majority of the agricultural workforce and are often compelled to work to meet their families' basic needs. While their contributions are recognized as central to the food and nutrition security of households and communities, their work is not recognized or supported adequately by public policy and social institutions. Women continue to face inequality across key development indicators including health, education and nutrition; discriminatory laws; and high levels of precarity in terms of income, employment conditions, safety and well-being. Lack of land titles is clearly a problem for women in terms of accessing credit and other resources, specifically from formal institutions such as banks and agricultural cooperatives. The lack of income in the hands of women agricultural workers has meant restrictions on dietary diversity and, in turn, nutrition security. A holistic strategy was proposed in the Draft Women Farmer's Entitlement Bill (2013) in India, which defined women as farmers, and sought equal entitlements to all resources and opportunities available to male farmers — whether land, water, credit or technology — alongside equal representation in decision-making around agricultural policies and programmes. Such legislations need to be adopted and enforced, to see India hunger free and nutritional secure (Rao, 2013). Yet laws are rarely enforced, and gender inequalities have deepened across sectors in many parts of the world, including South Asia

(Cornwall and Edwards, 2015). Violence against women is a structural barrier to the attainment of food and nutrition security via a range of pathways:

1. Withholding food or restricting funds to purchase food (Usta *et al.*, 2013);
2. Controlling when and how they eat (Lentz, 2018),
3. Pushing women into high-risk behaviour to secure money for food (Rao, 2019) or
4. The normalization of physical violence related to the non-performance of food-related work (food production, shopping, cooking *etc.*) (Bellows *et al.*, 2015; Chilton *et al.*, 2013). Gender inequity in India is a big challenge in attaining zero hunger as women are main players in Indian agriculture and their views, ideas and suggestions in agricultural farming are not being taken to board. Indian women manage agricultural work on family farms and receive no income, are usually overworked and have no wage-linked benefits (Rao, 2012). Apart from agricultural work, both household work and child and elder care responsibilities remain assigned to women within existing gender divisions of labour (Springer *et al.*, 2012), with the additional task of caring for their men. Agricultural work in the present context is unable to ensure adequate food, incomes or time for the accomplishment of caring and nurturing tasks, which are central to nutrition security. This has adverse impacts on women's own health and child and household nutrition (Kadiyala *et al.*, 2014).

## Possessions with India to become Hunger Free Country

We have discussed in length about the limitations India does confront in achieving zero hunger in the country. At the same time India has to identify the opportunities that can be explored and availed to see India food and nutritionally secure. Some of the opportunities that can be explored and researched for ameliorating the hunger in India include;

### Genetic Diversity

India has 12 per cent plant, 17 per cent animal and 10 per cent fish genetic sources of the world. This treasure with the India can be exploited to have the best capable, efficient, high producing and resilient varieties of crops and livestock to make India hunger free nation.

### Climatic Diversity

India has all the 15 climates of the world and 46 soil types that can be exploited to have food production in all the climatic and soil niches with nutritionally diverse foods. India is looking for designated areas with a specific type of food production depending upon the soil and climate conditions of that area. India has, long days and sunshine hours that is suitable for the cultivation round the year.

### Irrigation Potential

India has roughly half of the cultivable area rainfed without any proper mechanism of irrigation. India has scores of large rivers and water bodies that can be exploited and managed to irrigate the land for food production. This will also improve the utilization efficiency. Thus there exists great scope in increasing irrigated area, gross sown area and hence assured food production.

## Work Force

India has low age dependency ratio 49.2 per cent that means India has a good proportion of population in a workable age. This opportunity with the India is a great asset to face challenges. Today, India is one of the youngest country in the world with more than 62 per cent of the population in the working age group (15-59 years) and more than 54 per cent of the total population below 25 years of age but only 2 per cent are skilled (Newindia,301634)

## Livestock Resources

India is a cattle wealthy nation. Highest livestock population with great diversity makes India a potential country for animal products exportation. India has a long coastline that confers it a potential strength of economy through Blue Revolution. Milk, meat, fish and their products have a promising demand in many countries. India has to build up the export policies and quality control mechanism to invite exports of various food products.

## Gender Inequity

Indian rural women that plays a central role in agriculture farming, are less likely to be in a position to control productive resources necessary for agriculture in intensely patriarchal contexts and are more restricted in their access to and control over land, energy, water, pasture, forests, agricultural inputs, credit and insurance services, information, technology and markets (Food and Agriculture 2011). Despite progressive legislation in relation to inheritance of agricultural land in India and credit policies that target women's self-help groups across the subcontinent, however, enforcing rights, laws and policies, male hierarchy impedes its operationalization (Rao, 2017). Therefore, empowering women in its material resources shall bring a change in agriculture farming and food security.

## Welfare Measures

Although the national government has not strengthened the agricultural sector through better prices, assured procurement or investments in climate-proofing, several state governments in India have sought to ameliorate rural distress and other related effects of agrarian stagnation through welfare measures. Leaders in this respect are the states of Tamil Nadu and Andhra Pradesh with 55 and 48 welfare schemes, respectively.

Sanitation and health services: Reaching the goal of nutrition security is equally dependent on state provisioning of clean drinking water, sanitation and reliable health services. Sanitation, in the Clean India Campaign (Swachh Bharat Abhiyan), is interpreted as universal access to toilets. Although this is undoubtedly important, insufficient attention has been paid to the disposal of solid and liquid waste, contributing to the growing burden of infectious diseases (India State-level Disease Burden Initiative Collaborators, 2019). Similarly, whereas universal health insurance (Ayushman Bharat) may support curative care, primary health care provisioning (which is key to disease prevention) remains weak. The responsibility for both health and nutrition security is now placed on low-paid and poorly trained women community health and nutrition workers, with little power in their local, rural contexts (WHO, 2019). It is time to adequately support these workers to perform their roles, both through skill-building and the strengthening of referral and back-up health services. Health insurance need to be extended to maximum households in India.

## Way-Forward to have Hunger Free India

### Nutrition Sensitive Agriculture

Nutrition-sensitive agriculture is an approach that seeks to ensure the production of a variety of affordable, nutritious, culturally appropriate and safe foods in adequate quantity and quality to meet the dietary requirements of populations in a sustainable manner. Now when India is self sufficient in food production, the priority should be nutritionally rich and diverse foods. A large proportion of Indian population is nutritionally deficient in many essential macro and micro minerals. It is high time to use technologies like Biofortification, fortification, food diversification, soil enrichment, crop diversification that can lead to nutritional security. We need to have food plates decked with varieties of food items that can ensure nutritional demands of a person. Often food grain security is equated with food security, ignoring micronutrient deficiency and its harmful health hazards. There is need to expand consumption basket considerably to include non-grain items *viz.*, fruits, vegetables, beans and animal products. India is not adequately prepared to meet the requirements of dietary diversity, without enhanced trade. Diversification on a large scale (implementation at regional or national level) can help to enhance availability of diverse foods in markets and reduce prices of nutritious foods. Integrated farming systems (legume-based cropping systems including crop rotation and intercropping, rice-wheat farming systems) favour both diversification and sustainable intensification of production. Home gardening with emphasis on nutrient-dense varieties of vegetables and fruit trees and small-scale integrated farming systems (mixed crop-livestock-aquaculture systems) have potential to improve diet quality and raise levels of nutrition for producing households. The inclusion of animal source foods in the diet is an important food-based strategy for improving and safeguarding nutrition. In addition to protein and energy, animal source foods are excellent source of selected micronutrients like iron, zinc, calcium, vitamin A, vitamin B-12 and various essential amino-acids.

### Use of Neglected and Underutilized Species (NUS)

From a food system perspective, dietary and production diversity need to improve to address malnutrition. NUS are under-explored and can be called 'hidden treasures' that offer tremendous opportunities for fighting poverty, hunger and malnutrition. Director General FAO highlighted, the crucial role of NUS in the fight against hunger and are a key resource for agriculture and rural development (FAO, 2012a). Historically, underutilized plants have been used for food and other uses on a large scale and, in some countries, are still common, especially among small or marginal farmers in rural areas where many are traded locally, and few NUS are nutritious, climate resilient, economically viable and adapt to local conditions, especially in marginal areas. NUS have made their way to export niche markets around the world (Akinnifesi *et al.*, 2008). NUS have high nutritional value and can be an essential source of micronutrients, protein, energy and fibre, which contribute to food and nutrition security.

## Sustainable Intensification

There is little hope of increasing production from expanding the cultivated land area that otherwise is necessary to feed more than 9 billion population by 2050. More than 80 per cent of the necessary production increase will have to come from yield increases. Yet, the yield increase for all major food crops is declining, therefore, this challenge will not be met by continuing with the concept of conventional farming. There is no problem in the genetic potential of crops or production inputs, instead, the problem lies in the degradation of natural resources and their yield-related functions. Therefore, closing the yield gap, a different concept of sustainable intensification, has been coined by FAO (www.fao.org/ag/save-and-grow). Sustainable intensification means achieving the highest possible production, applying all necessary technologies, while keeping the environmental impact below the threshold of natural recovery. Conservation agriculture is based on key principles including minimum tillage, retaining a permanent soil cover, and crop diversification (Suraj and Behera 2014). Precision agriculture is an emerging example that utilizes data on spatial and temporal variations in the agroecosystem to make accurate agronomic decisions without wasting resources and time.

## Propagation of Indigenous Varieties

India has a rich diversification in crop varieties as well as soil and climate systems. Encouraging local production of diversified foods to be available locally can be a way out to eliminate hunger from the country. Up-gradation of local germplasm, identification and propagation of indigenous nutritious plant and animal species that are more resilient and adapted to local conditions can make India free from hunger. Selection and production of species and varieties should be based not only on yields but also on nutrient content (concept of nutrient productivity), thereby enhancing the nutrient supply of agricultural products, especially for micronutrients. Community-level initiatives for supporting the saving and exchange of seeds (community seed banks, village seed fairs, smallholder seed enterprises) and protecting ecosystems (community-based natural resource management, reforestation, promotion of micronutrient-rich forest foods) should be practised to enhance availability and access to genetic resources, Strengthening local food systems and empowering indigenous people in local production of nutritious, adapted and diversified cultivars. Locally available foods must find place in the food basket of its locals and its distribution to immediate areas to reduce wastage and cost.

## Accessibility to Food

Global food security is today more of an issue of access to food than its availability. Average dietary energy supply adequacy (ADESA) shows that minimum nutritional requirement per capita, is above sufficient levels globally (FAOSTAT) and yet, more than 800 million people across the globe are undernourished, and malnutrition. This implies grossly uneven distribution of food resources and emphasizes the urgent need for correcting distortions in food markets. It is a major issue that can be addressed by providing effective, efficient and economical public distribution system, transportation system, cooperative system and marketing system. Development of quality infrastructure for food storage, cold stores, cold chain facilities logistic services are of prime importance to ensure accessibility of population to food. Accessibility to food is linked to many aspects and a holistic approach is needed to ensure accessibility of population to food. It includes increasing purchasing capacity, reducing poverty, minimising food wastage, abolishing fragmented marketing with commission agents and middlemen mafia. One the key issues leading to food losses and wastage is that the production and consumption hubs are located in different regions across the entire country for example for three key commodities - potato, onion and tomato, which are core to Indian diet, the production and consumption hubs are far away, therefore, there is a need for efficient logistics and supply chain networks, both in terms of physical infrastructure and technology to manage and match the demand and supply to ensure accessibility of people to the food.

## Effective and Precision Irrigation System

50 per cent of the land is irrigated and 50 per cent remains rainfed in the country that leaves a good scope for increasing the food production through increasing irrigation. India does have great rivers with rich source as Himalayas that can be harvested for irrigation of farmlands. India needs to build infrastructure for its channelling to different areas across the States. Most parts of the India receive monsoon rains that can be harvested, stored and then distributed across the States to irrigate farmlands in the country. There has been little rather no change in arable land which was 155 million hectares in 1961 and 156 million hectares in 2016 but of course decrease in per capita holding from 0.34 hectares in 1961 to 0.12 hectares in 2016. Similarly India needs to develop crop varieties that have less water withdrawals.

## Sustainable Food Value Chains

Food systems encompass all the people, institutions and processes by which agricultural products are produced, processed, and brought to consumers (FAO SOFA 2013). "A food system gathers all the elements (environment, people, inputs, processes, infrastructures, institutions, *etc.*) and activities that relate to the production, processing, distribution, preparation and consumption of food, and the outputs of these activities, including socioeconomic and environmental outcomes" (HLPE -2014, p29). An effective food system transformation will necessitate firm commitments at the international and national levels to promote the required policies and investments at national and local levels. In practical terms, this would need to include four main dimensions: (i) underpinning healthier populations by enabling access to nutritious and healthy food for all; (ii) guaranteeing sustainable food production, processing, trade, and retailing; (iii) mitigating and adapting to climate change; (iv) improving smallholder farmer livelihoods and resilience by enhancing prosperity in farming and rural communities. FAO has developed guiding principles on sustainable food value chains which can be used as a framework for upgrading value chains (FAO- 2014b).

## Minimising Wastages

Wastages do happen from food production to consumer's plate and this loss if reduced can alone make India free from hunger. Use of efficient strategies at the farm level, efficient transportation system from farm to stores and consumers and ultimately proper utilization by the consumers without wasting much can do wonders in eliminating hunger from the country. In India food losses occur mainly at early stages of the food value chain. Strengthening the supply chain through the direct support of farmers and investments in cold chain infrastructure, transportation, and safe packaging could help to reduce the amount of food loss. Post-harvest handling, processing and storage contributes to a secure food supply and thus of nutrients throughout the year. It also reserves the quality of harvested raw material as it moves along the food supply chain from the producer to the market; reduces losses; and makes fresh produce available in local markets as well as in distant locations. There are some sensitive groups which collect the surplus food left on the table and distribute them among the poor. India need to sensitise its high income consumers to judicially order foods in restaurants, gatherings and parties so that wastage is minimised.

## Food Processing and Value Addition of Food Products

Out of the total agricultural output, the share of food processing sector in India was only 46.87 per cent in 2017-18. Comparatively, the share for other developing countries such as Brazil (70 per cent), Malaysia (80 per cent) and Philippines (78 per cent) is much higher (Reserve Bank of India, RBI-2020). One of the effective methods to preserve the surplus food particularly perishable items and make it available to the consumers in the country or outside the country is to process the food. Value addition is one of the effective strategies to preserve the foods, to utilize the non-conventional foods and to fetch more. The strategy can be effectively utilized to make use of agriculture by-products, which is only 25 per cent.

## Research and Innovations

Research and innovations will always remain effective tools in any system to face the challenges. The nations which are using innovative techniques are economically and technologically ahead. In India we have our own demands in agriculture farming system. Our landholdings are very little and thus need to have farm machinery and technologies that favour our small farm lands be it in seed sowing, land tilling, crop planting, thrashing, chopping, harvesting *etc.* There is need for more investment on applied post-harvest management research/technologies, policy change to focus on infrastructure development including cold storage, refrigerated transportation, and efficient distribution systems *etc.*

Research is needed to have new crop cultivars (varieties and hybrids) with improved input use efficiency such as water/nutrient use efficiency, better light interception, drought tolerance, tolerance/resistance to insects and diseases and competitive ability against weeds, plasticity towards unpredictable weather elements and better nutritional quality per unit of produce. There is urgent need to progress in technologies such as remote sensing, weather monitoring, and drones endowed with hyperspectral imaging cameras *etc.* to collect data to develop timely predictions to support decision making processes of growers related to input allocation, pest management and other production practices.

## Adoption of Modern Technologies and Data Analytics

In any system it is the data analysis that provides the leads for the success of technology. Adoption of technology is must and our agriculture extension wings, KVKs are to be strengthened to take the technologies from laboratory to the field and feed back to the laboratories about its impact in increasing food production. Extension services are of paramount importance and KVKs are the connecting links between laboratories and fields and good liaison between the two can really help in getting India out of hunger.

## Awareness among Consumers

Awareness is must in present scenario of globalization. Consumer is presented with varieties of food items claiming lot many nutritional benefits with high pricing for the product. There is need to aware the consumer through target campaigning. There is need to educate the consumers about the requirement of different varieties in the food basket according to the physiological status, age, sex and physical activities of the consumers. Pregnant ladies, lactating mothers, neonates, juveniles, athletes, women and men have different food requirements and they must be made aware of their nutritional needs and food varieties that can fulfil their nutritional requirements so that a healthy society is ensured.

## Strict Implementation of Product Standard Specifications and Food Safety Standards

India has a huge potential of export of agricultural and livestock products but has not achieved much progress in it simply because of lack of standard product specification and certification. Quality control is a major issue with the products that have an export market. India has not been able to reach its full export potential due to fragmented supply chain and gaps in mitigating the demand of importing countries (Mukherjee *et al.,* 2019; Goyal *et al.,* 2017). Specifically, many of the developed countries demand full product traceability from farm to market. The issues faced by our exporters include frequent rejection of products for not meeting the MRLs for pesticides, high presence of harmful organisms in the plant produce, pest infestation in fresh food products such as mango and vegetables, lack of harmonisation of standards, not fulfilling export requirements of advanced countries with respect to technology used or laboratory testing procedures, among others (Mukherjee *et al.,* 2019). Therefore India has to ensure quality control of agricultural products through establishment of high end analytical laboratories, sanitation procedures, SPS, traceability and observance of standard safety measures that can boost the export of food items from India in the global market and make its economy flourishing.

## Regional Trade Cooperation and Liberalization in Agriculture Products

India is the biggest partner in SAARC, however, the regional trade cooperation between the countries is very meagre may be because of undue regulatory restrictions, procedural and infrastructural barriers to trade or due to certain political reasons. While creation of liberalized trade regimes through trade agreements dealing with both tariff and non-tariff barriers are essential aspect of regional cooperation, its scope goes beyond the ambit of trade. India needs to strengthen the trade agreements with neighbouring countries to make diversified food available to the population at reasonable and affordable rates. Unfair trade agreements and subsidies create market access and price advantages for enterprises from the industrial nations. Developing countries like India primarily export raw materials, the profits are skimmed off by rich states. Thus India needs strong liberalized trade policies, political will and firm resolve. India has to increase intra-regional agriculture trade. The recently introduced Food Security Standard (FSS) should manage food and nutrition security with greater justice worldwide. In terms of facilitating availability, open trade; (a) allows flow of food resources to flow from surplus to deficit markets, (b) enables dietary diversity by increasing the choices of food products available for consumption (important for nutritional fulfilment), (c) enhances marketability of agricultural outputs and thereby supports sustenance of food production, and (d) helps to reduce wastage and improve efficiency of food distribution. By way of improving accessibility, open trade; (a) helps to curb food price inflation and thereby support price stability and affordability, and (b) enables market expansion, increased export earnings, and supports livelihoods and purchasing power of farm sector dependents (Nagesh Kumar and Joseph George, 2020).

# REFERENCES

Akinnifesi. F.K., Leakey, R.R.B., Ajayi, O.C., Sileshi, G., Tchoundjeu, Z., Matakala, P. and Appu, P. S. 1996. Land Reforms in India: A Survey of Policy, Legislation and Implementation. Vikas Publishing House, New Delhi.

Balarajan. Y., S. Selvaraj and S.V. Subramanian. 2011. Health care and equity in India.The Lancet. 377: 9764, pp 505-515. https: //doi.org/10.1016/S0140-6736(10)61894-6

Basu Kaushik and Annemie Maertens. 2008. The Oxford Companion to Economics in India.

Bellows.A. C., Lemke. S., Jenderedjian. A. and Scherbaum. V. 2015. Violence as an unrecognized barrier to women's realization of their right to adequate food and nutrition: case studies from Georgia and South Africa. Violence Against Women. 21, 1194–1217.

Besley.T and Burgess. R. 2000. "Land Reform, Poverty Reduction, and Growth: Evidence from India." The Quarterly Journal of Economics 115: 389-430.

Bhatia. B. M. 1985. "Food Security in South Asia", Oxford and IBH Publishing Co. New Delhi.

Cavatorta. E., Shankar. B. and Flores-Martinez. A. 2015. Explaining cross-state disparities in child nutrition in rural India. World Dev. 76, 216–237.

Chatterjee. K., Sinha. R.K., Kundu. A.K., Shankar. D., Gope. R., Nair. N. and Tripathy. P.K. 2016. Social Determinants of Inequities in Under-Nutrition (Weight-For-Age) among Under-5 Children: A Cross Sectional Study in Gumla District of Jharkhand, India. International Journal for Equity in Health, 15: 104.

Chilton. M. M., Rabinowich. J. R. and Woolf. N. H. 2013. Very low food security in the USA is linked with exposure to violence. Public Health Nutr. 17, 73–82.

Comprehensive National Nutrition Survey (CNNS) National Report (Ministry of Health and Family Welfare, Government of India, UNICEF and Population Council, 2019.

Cornwall. A. and Edwards. J. 2015. Introduction: Beijing+20—where now for gender equality? IDS Bull. 46, 1–8.

Devereux Stephen. 2000. Famine in the twentieth century (PDF) (Technical report) IDS Working Paper 105. Brighton: Institute of Development Studies.

Dutta. M., Selvamani. Y., Singh. P., and Prashad. L. 2019. The Double Burden of Malnutrition among Adults in India: Evidence from the National Family Health Survey-4 (2015-16). Epidemiol Health. 41.

FAO. 2020. India at a glance. http: //www.fao.org/india/fao-in-india/india-at-a-glance/en/#: ~: text=India per cent 20is per cent 20the per cent 20world's per cent 20largest,poultry per cent 2C per cent 20livestock per cent 20 and per cent 20plantation per cent 20crops.

FAO. 2017. The Future of Food and Agriculture, Trends and Challenges. Rome, 2017.

FAO. 2015. Natural Capital Impacts in Agriculture. FAO.Rome

FAO. 2014b. Developing sustainable food value chains – Guiding principles. FAO. Rome

FAO. 2012a. Neglected crops need a rethink – can help world face the food security challenges of the future. (www.fao.org/news/story/en/item/166368/icode/).

FAO. 2011. The State of Food and Agriculture 2010–2011: Women in Agriculture—Closing the Gender Gap for Development.

Ford. N.D., Patel. S.A., and Narayan. K.M.V. 2017. Obesity in Low-and Middle-Income Countries: Burden, Drivers, and Emerging Challenges. Annual Review of Public Health, 38, pp. 145-164. https: //www.annualreviews.org/doi/full/10.1146/annurev-publhealth-031816-044604.

Global Hunger Index (GHI, 2019). https: //www. Globalhungerindex.org

Global Nutrition Report. 2020. Action on Equity to End Malnutrition. https: // globalnutrition report.org/reports/2020-global-nutrition-report/

Gokarn. S., and Kuthambalayan. T.S. 2017. Analysis of Challenge Inhibiting the Reduction of Waste in Food Supply Chain. United States Department of Agriculture. https: //pubag.nal.usda.gov/catalog/5825414

Gope. R. K. *et al.,* 2019. Effects of participatory learning and action with women's groups, counselling through home visits and crèches on undernutrition among children under three years in eastern India: a quasi-experimental study. BMC Public Health 19, 962.

Goyal. T.M., Mukherjee. A., and Kapoor. A. 2017. India's Export of Food Products: Food and Safety Related Issues and Way Forward. Indian Council of International Economic Relations (ICRIER) Working Paper, No. 345.

Gulati. S. and Misra. A. 2014. Sugar Intake, Obesity, and Diabetes in India. Nutrients, 6 (12): 5955–5974.

India State-level Disease Burden Initiative Collaborators. 2019. The burden of child and maternal malnutrition and trends in its indicators in the states of India: the Global Burden of Disease Study 1990–2017. Lancet Child Adolesc. Health 3, 855–870.

Kadiyala. S., Harris. J., Headey. D., Yosef. S. and Gillespie. S. 2014. Agriculture and nutrition in India: mapping evidence to pathways. Ann. N. Y. Acad. Sci. 1331, 43–56.

Kwesiga. F.R. 2008. Indigenous Fruit Trees in the Tropics: Domestication, Utilization and Commercialization. CAB International Publishing, Wallingford, UK, pp. 438.

Lentz, E. C. 2018. Complicating narratives of women's food and nutrition insecurity: domestic violence in rural Bangladesh. World Dev. 104, 271–280.

Maharatna. Arup. 1996. The Demography of Famines: an Indian Historical Perspective. Oxford University Press. ISBN 978-0-19-563711-3.

Ministry of Statistics and Programme Implementation (MOSPI) and World Food Programme (WFP). 2019. Food and Nutrition Security Analysis, India, 2019. Government of India. http: //www.indiaenvironmentportal.org. in/files/ file/Food per cent 20 and per cent 20Nutrition per cent 20Security per cent 20Analysis.pdf

Mukherjee. A., Goyal. T.M., Miglani. S., Kapoor. A. 2019. SPS Barriers to India's Agriculture Exports. Learning from the EU Experiences in SPS and Food Safety Standards. ICRIER Report. Available at https: //icrier.org/pdf/SPS_Barriers_ to_ India_Agriculture_ Export.pdf

Nagesh Kumar and Joseph George. 2020. Regional Cooperation for Sustainable Food Security in South Asia. Published by Routledge India. Pp 20 -280

Negi. S. and Anand. N. 2017. Factors Leading to Losses and Wastage in the Supply Chain of Fruits and Vegetables Sector in India. Chapter in Energy Infrastructure and Transportation Challenge and Way Forward. https: //www.researchgate. net/profile/Saurav _Negi/publication/298891316_Factors_Leading_ to _Losses_and_ wastage_in_the_Supply_Chain_ of_Fruits_and_ Vegetables_

Sector_in_ India/links/56ed3dcc08ae4b8b5e 73eea4/Factors-Leading-to-Losses-and-Wastage-in-the-Supply-Chain-of-Fruits-and-Vegetables-Sector-in-India.pdf

Newindia. 2017. https://www.mygov.in/newindia/wp-content/uploads/2017/09/newindia 301634_1504694260.pdf

Nguyen. H.A. 2019. Under-nutrition During Pregnancy. IntechOpen. https://www.intechopen.com/books/complications-of-pregnancy/undernutrition-during-pregnancy

Nie, P., Rammohan, A., Gwozdz, W., and Sousa, P. 2016. Developments in Undernutrition in Indian Children Under Five: A Decompositional Analysis. Discussion Paper Series, No 9893, IZA. http://ftp.iza.org/dp9893.pdf, Oxford University Press, New Delhi.

Pingali. P. 2012. Green Revolution: Impacts, limits, and the path ahead. PNAS 109 (31): 12302-12308.

Pingali. P.L., Aiyar. A., and Abraham, M. 2019. Transforming Food Systems for Rising India. Palgrave Studies in Agricultural Economics and Food Policy. https://www. researchgate. net/publication/331500711_Transforming_Food_Systems_for _a _Rising_India

Prasad, V., Sinha, D., Chatterjee, P. and Gope, R. K. 2018. Outcomes of children with severe acute malnutrition in a tribal day-care setting. Ind. Pediatr. 55, 134–136.

Rao, N. 2019. From abandonment to autonomy: gendered strategies for coping with climate change, Isiolo country, Kenya. Geoforum 102, 27–37.

Rao, N. 2012. Male 'providers' and female 'housewives: a gendered co-performance in rural North India. Dev. Change 43, 1025–1048.

Rao, N. 2013. Rights, recognition and rape. Econ. Polit. Weekly 48, 18–20.

Raykar N., Majumdar M, Laxminarayan R., Menon P. 2015. India Health Report: Nutrition 2015. New Delhi, India: Public Health Foundation of India.

Report on Fifth Annual Employment - Unemployment Survey. 2015-16. Government of India Ministry Of Labour and Employment. http://labourbureaunew.gov.in/UserContent/EUS_5th_1.pdf

Singh, A. 2020. Childhood Malnutrition in India. IntechOpen. http://dx.doi.org/10.5772 intechopen.89701

Smith, L. C. and Haddad, L. 2015. Reducing child undernutrition: past drivers and priorities for the post- MDG era. World Dev. 68, 180–204.

Springer, K. W., Hankivsky, O. and Bates, L. M. 2012. Gender and health: relational, intersectional, and biosocial approaches. Soc. Sci. Med. 74, 1661–1666.

Suraj. B. and Behera. U.K. 2014. Conservation agriculture in India – Problems, prospects, and policy issues. International Soil and Water Conservation Research, 2: 1-12

How to feed the world 2050. http: //www.fao.org/fileadmin/templates/wsfs/ docs/expert_paper/How_ to_ Feed_the_World_in_2050.pdf

Swagata Yadavar. 2017. Factchecker.in reported in December 2017.

Thorner Daniel. 1976. Agrarian Prospect in India. New Delhi: Allied Publishers.

Thow. A.M., Kadiyala. S., Khandelwal. S., Menon. P., Downs. S. and

Reddy. K.S. 2016. Toward Food Policy for the Dual Burden of Malnutrition: An Exploratory Policy Space Analysis in India. Food and Nutrition Bulletin, 37: 3.

United Nations. 2015 "Transforming our world: the 2030 Agenda for Sustainable Development". United Nations – Sustainable Development knowledge platform. https: //sustainable development.un.org/post2015/transforming ourworld

United Nations (UN). 2019. World Population Prospects 2019. Department of Economic and Social Affairs, Population Division https: //population.un.org/ wpp/Publications/Files/WPP2019_Highlights.pdf

UNICEF (United Nations International Children's Education Fund). 2020. Malnutrition Prevalence remains Alarming: Stunting is Declining too Slowly while Wasting Still Impacts the Lives of far too Many Young Children. Available at https: //data.unicef.org/topic/nutrition/malnutrition/

Usta. J., Makarem. N. and Habib. R. 2013. Economic abuse in Lebanon: experiences and perceptions. Violence Against Wom. 19, 356–375.

Vipul Vivek. 2017. IndiaSpend reported in May 2017

WHO. 2019. Delivered by Women, Led by Men: A Gender and Equity Analysis of the Global Health and Social Workforce Human Resources for Health Observer Series No. 24.

WHO (World Health Organization). 2018. Obesity and Overweight.: https: //www. who.int/en/news-room/factsheets/detail/obesity -and-overweight

World Integrated Trade Statistics (WITS). 2018. <https: //wits. worldbank.org/ Country Profile/en/Country/IND/Year/2018>

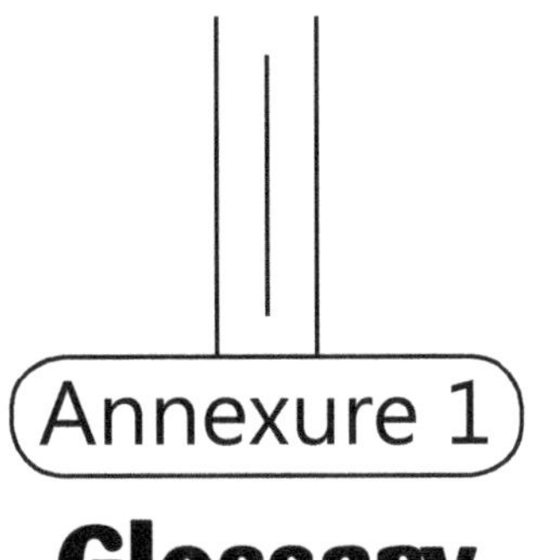



# Glossary

**1000 days – Window of opportunity:** The period between conception and two years of age when irreversible damage caused by malnutrition can and should be prevented. Reference: The 2008 Lancet Series on Maternal and Child Undernutrition

**Acute malnutrition (wasting/low weight-for-height):** Wasting or thinness indicates in most cases a recent and severe process of weight loss, which is often associated with acute starvation and/or severe disease. Children under 5 years of age are the most exposed to risks of acute malnutrition, in particular when transitioning from exclusive breastfeeding to complementary feeding. Reference: WHO

**Aquaponics:** Aquaponics refers to any system that combines conventional aquaculture (raising aquatic animals such as snails, fish, crayfish or prawns) with hydroponics (the cultivation of plants by placing the roots in liquid nutrient solutions rather than in soil) in a symbiotic environment, where the plants metabolize the by-products from aquaculture, keeping the water environment clean for aquatic animal life. Reference: FAO

**Balanced diet:** A diet that provides an adequate amount and variety of food to meet a person's macro and micronutrient needs for a healthy, active life.

**Bioavailability:** The amount of an ingested nutrient that can be digested, absorbed and used by the body. Reference: Codex Alimentarius

**Bottom of the Pyramid:** The Bottom of the pyramid (BoP) refer to the bottom of the wealth pyramid, which is the largest but also the poorest socio-economic group. BoP models refer to business models targeted at providing goods and services to the poorest, through product innovation (e.g. small packaging, to respond to the need of those with little purchasing power) or process innovation (e.g. franchising with small retailers which serve the poor).

**Chronic malnutrition (stunting/low height-for-age):** A form of growth failure that causes both physical and cognitive delays in growth and development, which arises when the body is not able to absorb the sufficient amounts of nutrients (due to lack of access to adequate foods and/or to disease) to meet dietary energy and nutrient requirements over a prolonged period of time. Reference: WHO

**Climate-smart agriculture (CSA):** An approach that helps to guide actions needed to transform and reorient agricultural systems to effectively support development and ensure food security in a changing climate. CSA aims to tackle three main objectives: sustainably increasing agricultural productivity and incomes; adapting and building resilience to climate change; and reducing and/or removing greenhouse gas emissions, where possible. Reference: FAO

**Complementary feeding:** Nourishment of an infant with foods in addition to breastmilk or breastmilk substitutes. After six months of age, when breastmilk is no longer enough to meet the nutritional needs of the infant, complementary foods should be added to the diet of the child. Reference: FAO

**Dietary Diversity:** A measure of the variety of foods from different food groups consumed by an individual or by a group over a determined period. Reference: FAO

**Double burden of malnutrition:** The co-existence of undernutrition (wasting, stunting and micronutrient deficiencies) along with overweight/obesity, within an individual, household or group, across the life course. Reference: FAO/WHO

**Enabling environment (for food security and nutrition):** The enabling environment for food security and nutrition comprises commitments and capacities across a range of dimensions such as policies, programmes and legal frameworks; mobilization of human and financial resources; coordination mechanisms and partnerships; and evidence-based decision making. Reference: FAO

**Energy-dense food:** A food with a high content of calories (energy) per gram. Highly-processed energy-dense foods that are high in sugars, saturated fats and/or salts but poor in micronutrients are likely to adversely affect health. Reference: FAO/WHO

**Food environment:** The food environment is one of the emerging concepts associated with food systems and nutrition. It designates the interface between the food system and consumers. The food environment is defined as the availability, affordability, convenience and desirability of various foods. The food environment is directly affected by the food system, and in turn affects diet quality and nutritional status. In research, the concept of food environment has been mainly used in relation to dietary quality issues in high-income countries (i.e. overweight, obesity and Non-Communicable Diseases [NCDs]). Reference: Herforth and Ahmed, 2015

**Food system:** A food system encompasses all the people, institutions and processes by which agricultural products are produced, processed and brought to consumers. A food system gathers all the elements (environment, people, inputs, processes, infrastructures, institutions, *etc.*) and activities that relate to the production, processing, distribution, preparation and consumption of food, and the outputs of these activities, including socioeconomic and environmental outcomes. References: SOFA 2013, HLPE 2014

**Food-based approach:** An approach which recognizes the central role of food for improving nutritional status. A food-based approach recognizes the multiple benefits (nutritional, physiological, mental, economic, social and cultural) that come from enjoying a variety of foods. Food-based approaches can be complemented with strategies that rely on medically-based interventions such as vitamin and mineral supplementation. Reference: FAO

**Food-based dietary guidelines:** Food-based dietary guidelines (also known as dietary guidelines) are intended to establish a basis for public policies, programmes and actions fostering healthy eating habits and lifestyles. They provide advice on foods, food groups and dietary patterns to promote overall health and prevent chronic diseases. Reference: FAO

**Healthy diets:** Diets that provide protection against malnutrition in all its forms, as well as non-communicable diseases (NCDs), including diabetes, heart disease, stroke and cancer. Reference: WHO

**Home Grown School Meals (HGSM):** Home grown school meals (HGSM) is a school meals model to provide school children with safe, diverse and nutritious food, sourced from local smallholders. Reference: WFP

**Indigenous crop:** Native plants to a given area in geologic time. This includes neglected and underutilized species. Reference: FAO

**Integrated multi-trophic aquaculture (IMTA):** A method of aquaculture whereby different species are raised together in a given controlled area, connected by nutrient and energy transfer through water (e.g. recycling of by-products from a given species serves as feed for another). Reference: FAO.

**Livelihood:** A livelihood comprises the capabilities, assets (natural, human, physical and financial) and activities required for a means of living: a livelihood is sustainable which can cope with and recover from stress and shocks, maintain or enhance its capabilities and assets, and provide sustainable livelihood opportunities for the next generation; and which contributes net benefits to other livelihoods at the local and global levels and in the long and short term. Reference: R. Chambers and G. Conway, 1992

**Low birth weight:** Weight at birth of less than 2 500 grams, contributing to a range of poor health outcomes and an increased mortality risk. Reference: WHO

**Malnutrition:** An abnormal physiological condition caused by deficiencies, excesses or imbalances in energy and/or nutrients necessary for an active, healthy life. Malnutrition includes undernutrition, micronutrient deficiencies, overweight and obesity, conditions that can arise separately or coexist. Reference: FAO/WHO

**Micronutrient deficiency (hidden hunger):** Lack of vitamins, minerals and/or trace elements which are essential for the proper functioning, growth and metabolism of a living organism. Usually caused by poor diets, it is often referred to as "hidden hunger" as its physical symptoms are not obvious to detect, while its consequences can be deadly. It can, and often does, coexist with undernutrition and overweigh and obesity. Reference: FAO/WHO

**Neglected and underutilized species (NUS):** Neglected and underutilized species (NUS) are those to which little attention is paid or which are ignored by agricultural researchers, plant breeders and policymakers. Typically, NUS are not traded as commodities. They are wild or semi-domesticated varieties and non-timber forest species adapted to particular, often quite local, environments. Many of these varieties and species, along with a wealth of traditional knowledge about their cultivation and use, are being lost at an alarming rate. Reference: Bioversity International

**Non-Communicable Diseases (NCDs):** Non-communicable diseases (NCDs), also known as chronic diseases, are not passed from person to person. They are of long duration and generally slow progression. The 4 main types of non-communicable diseases are cardiovascular diseases (like heart attacks and stroke), cancers, chronic respiratory diseases (such as chronic obstructive pulmonary disease and asthma) and diabetes. Reference: WHO

**Nutrient-dense crop:** Crops with a high content of nutrients per gram. Reference: FAO

**Nutrient productivity:** This is a measure to assess the extent to which the agriculture production is meeting the nutritional needs of its population for 9 nutrients: energy, protein, dietary fiber, iron, zinc, calcium, vitamin A, vitamin C and folate. It combines yield, the nutrient composition of the agricultural product and the nutrient requirement for these 9 nutrients for humans. Reference: FAO

**Nutrient-rich foods:** Foods with a high content of nutrients per gram. Reference: FAO/WHO

**Nutrition:** Nutrition is the intake of food, considered in relation to the body's dietary needs. Good nutrition - an adequate, well balanced diet combined with regular physical activity - is a cornerstone of good health. Poor nutrition can lead to reduced immunity, increased susceptibility to disease, impaired physical and mental development, and reduced productivity. Reference: FAO/WHO

**Nutrition-sensitive interventions:** Interventions in any sector, which do not necessarily have nutrition as predominant goal but are designed to also address some of the underlying causes of malnutrition (which include household food security, care for mothers and children, and primary health care services and sanitation). Reference: Lancet 2008/FAO

**Nutrition-specific interventions:** Interventions which predominant goal is nutrition, designed primarily to address immediate determinants of malnutrition such as adequate food and nutrient intake, treatment of acute malnutrition, care-giving practices and reducing the burden of infectious diseases. Reference: Lancet 2008/WHO

**Nutrition transition:** "Nutrition transition" refers to a trend of decreased proportion of undernourished children and adults, and a rise in the proportion of children and adults who are overweight, obese and suffer from diet related non-communicable diseases. It is the outcome of a dietary transition from traditional diets, characterized by high intakes of cereals and fibers, to affluent diets, rich in saturated fats, sugar and highly processed foods as well as of broader lifestyle changes.

**Overweight and obesity:** Body weight that is above normal for height, as a result of an excessive accumulation of fat. A Body Mass Index (BMI) comprised between 25 and 30 corresponds to Overweight. A BMI above 30 corresponds to Obesity. BMI is defined as the body mass (expressed in kilogrammes) divided by the square of the body height (expressed in meters). Reference: WHO

**Processed food:** According to the extent of processing being used, foods can be distinguished in: - Unprocessed foods: food consumed shortly after harvesting, slaughtering *etc*.- Processed culinary ingredients: food products extracted and refined from constituents of foods (e.g. plant oils, animal fats, starches, sugar) and salt.- Minimally processed foods: unprocessed foods altered in ways that do not add or introduce any substance, but that may involve subtracting parts of the food (e.g. through cleaning, peeling, squeezing, filleting, drying, pasteurization and freezing, *etc*.).- Processed foods: food made by adding a culinary ingredient to unprocessed or minimally processed foods; the resulting product retain the basic identity and most of the constituents of the original food, but the substances added infiltrate the foods and alter their nature (e.g. canned or bottled vegetables or legumes, tinned fish preserved in oil, bread made from cereal flours, water, ferments and salt).- Ultra-processed foods and drink products: products which are formulated mostly or entirely from substances derived from foods, with little or even no whole food content. They typically contain various combinations of preservatives; stabilizers, emulsifiers, solvents, sweeteners, colours, flavours *etc*. (e.g. mass-manufactured breads and pastries; confectionery; ready meals, canned or dehydrated soups; chips; snacks; sugared or sweetened drinks, *etc*.). Reference: NOVA Food definition and classification system

**Ready-to-eat food:** Any food that is normally eaten in its raw state or any food handled, processed, mixed, cooked, or otherwise prepared into a form, which is normally eaten without further virucidal steps, e.g. by processing. Reference: FAO

**Ready-to-use supplementary food:** Ready-to-use supplementary food is a type of ready-to-use food that is specifically designed for the treatment of moderate acute malnutrition in children 6-59 months of age. RUSFs are fortified with micronutrients and contain essential fatty acids and quality protein to ensure a child's nutritional needs are met. Reference: FAO

**Ready-to-use therapeutic food:** Ready-to-use therapeutic food has revolutionized the treatment of severe malnutrition – providing foods that are safe to use at home and ensure rapid weight gain in severely malnourished children. Reference: FAO

**Social marketing:** Social marketing seeks to alter specific behaviour practices (in this case, dietary practices) in pursuit of a social good (in this case, improved nutrition) by drawing upon commercial marketing methods. Reference: FAO

**Social protection:** Social protection encompasses initiatives that provide cash or in-kind transfers to the poor, protect the vulnerable against risks and enhance the social status and rights of the marginalized; all with the overall goal of reducing poverty and economic and social vulnerability. Reference: FAO

**Staple food:** A staple food is one that is eaten regularly and in such quantities as to constitute the dominant part of the diet and supply a major proportion of energy and nutrient needs. A staple food does not meet a population's total nutritional needs: a variety of foods is required. Reference: FAO

**Sustainable diets:** Sustainable Diets are those diets with low environmental impacts which contribute to food and nutrition security and to healthy life for present and future generations. Sustainable diets are protective and respectful of biodiversity and ecosystems, culturally acceptable, accessible, economically fair and affordable; nutritionally adequate, safe and healthy; while optimizing natural and human resources. Reference: FAO/Biodiversity International

**Undernutrition:** The outcome of insufficient intake, and/or poor absorption and/ or poor biological use of nutrients consumed as a result of repeated infectious disease. It includes being underweight for one's age, too short for one's age (stunted), dangerously thin for one's height (wasted) and deficient in vitamins and minerals (micronutrient malnutrition). Reference: FAO

**VAC systems:** VAC (Vuon, Ao, Chuong) are small scale, integrated systems used in Vietnam and typically include crop farming (*e.g.* staple production and home gardening for fruits and vegetables), aquaculture (*e.g.* flooded rice paddies used as fish ponds) and animal husbandry (*e.g.* small poultry, which also provide fertilisers for crop production). Reference: FAO

# Index

Accessibility: 19, 23, 39, 103, 178, 252

Adaptation: 9, 63, 128, 130, 132-134, 193, 216,

Biodiversity: 2, 17, 24, 30, 31, 39, 53-56, 63, 68, 133, 193, 212, 216,

Biofortification: 17, 113, 115-119, 250

Breeding: 9, 114-116, 119, 133, 169, 170, 172, 173, 176, 178, 194, 216, 217, 220, 231

Challenges: 2, 3, 8, 30, 38, 49, 53, 80, 86, 89, 104-107, 111, 114, 118, 127, 130, 139, 157, 170, 173, 178, 181, 199, 207, 208, 212, 214, 220, 249, 253

Climate change: 1, 3, 4, 5, 8, 10, 11, 30, 31, 32, 37, 38, 39, 40, 49, 55, 56, 105, 108,109, 111, 118, 127, 128,130, 133, 134, 179, 244, 245

Climate resilient: 1, 4, 10, 130, 133, 134, 235, 251

Cold chain: 23, 24, 71, 90, 173, 215, 246, 252, 253.

Conservation: 6, 9, 40, 41, 53, 57, 62, 110, 130, 133, 134, 171, 173, 180, 251.

Deficiency: 5, 17, 18, 72, 73, 97, 105, 113, 114, 205, 247, 250.

Diversification: 3, 4, 6, 10-12, 16, 24, 30, 110, 111, 114, 118, 120, 127, 193, 207, 209, 250-251.

Ecology: 7, 39, 60, 78, 191

Economical: 3, 4, 7, 10, 116, 170, 179, 202, 254.

Environment: 4, 8-10, 12, 15, 16, 19, 22, 24-26, 30, 31, 33, 54, 55, 73, 110, 116, 128, 134, 148, 153, 157, 159, 184, 191, 193, 194, 206, 218, 227, 252.

FAO: 4-7, 11, 15, 30-32, 71, 78, 109, 207, 239, 250, 252.

Food fortification: 18, 114, 115

Food Insecurity: 2, 11, 79, 104, 219

Food loss: 22, 23, 26, 78, 83, 109, 139, 252, 253

Food production: 3, 4, 6, 16, 20, 24, 30, 31, 39, 40, 44, 53, 67, 68, 78, 81, 85, 90, 103, 107, 109, 139, 191, 197, 211, 212, 223, 235, 239, 240, 244, 248, 249, 252, 253.

Food Security: 252, 255, 3, 10, 15,20, 26, 30, 33, 39, 45-49, 63, 67, 79, 103-105, 107-109, 128, 139, 152, 153, 155- 157, 165, 179, 184, 197, 205, 223, 235.

Food Systems: 1- 4, 11, 12, 15-20, 24, 30-33, 45, 119, 139, 140, 193, 251, 252, 262

Food Wastage: 80, 109, 252,

Future Smart Food: 1, 7, 8, 10, 11, 250

Gender equality: 22, 26, 37, 43, 48, 49

Global: 1, 3, 6, 8, 16, 19, 33, 37, 39, 43, 45, 53- 55, 60, 62, 63, 68, 69, 72, 74, 87, 89, 95, 97, 98, 100, 103, 104, 109, 113, 127, 128, 143, 157, 158, 165, 169, 184, 189, 206, 213, 219, 225, 227, 228, 231, 237, 241, 246, 252, 255,

Indian population: 39, 235, 250

Intensification: 3, 10, 12, 16, 53, 54, 57, 105, 127, 193, 194, 245, 250, 251.

Malnutrition: 2-5, 11, 12, 16-19, 21, 24, 29, 49, 71-76, 79, 86, 88, 96, 98-103, 114, 116, 120, 133, 153, 159, 165, 206, 223, 224, 243, 244, 247, 250.

Mechanization: 9, 45, 105, 217

Micronutrients: 4, 16, 40, 97, 100, 114-116, 205, 245, 250

Natural resources: 3, 8, 39, 108, 109, 127, 244,

NUS: 1, 2, 7, 8, 11

Nutrition sensitive: 15, 114, 153, 235, 250

Obesity: 1, 10, 11, 19, 30, 71-74, 86, 89, 96, 114, 159, 244.

Opportunities: 6, 8, 11, 20, 22, 44, 46, 47, 53, 69, 70, 159, 165, 193, 199, 207, 212, 225, 247, 48, 250.

Pesticide: 23, 55, 56, 57, 61, 185, 213,

Policies: 4, 6, 12, 18, 19, 22, 24-26, 30, 31, 41, 47, 49, 62, 68, 80, 87, 89, 90, 104, 115, 134, 144, 145, 146, 155, 181, 217, 246, 247

Practices: 6, 19, 23, 25, 31, 37, 41, 44, 54-57, 60, 61, 67, 71, 97, 101, 102, 104, 108, 110, 115, 117, 128, 130, 134, 141, 148, 153, 157, 172, 179, 180, 190, 220, 221, 226, 228, 231, 243, 254.

Public sector: 19, 24, 33.

Staple crops: 2-4, 11, 115

Strategies: 8, 10, 19, 22, 24, 41, 89, 90, 101, 106, 109, 128, 130, 133, 34, 145, 146, 153-55, 157. 158, 219, 227, 253.

Stunting: 30, 45, 71, 88, 96-98, 114, 144-146, 152-159, 243, 247,

Sustainability: 1, 9, 11, 12, 26, 30, 31, 33, 40, 54, 55, 63, 105, 109, 111, 139, 140, 179, 209, 213, 220.

Sustainable development goal: 1, 3, 30, 37, 54, 57, 79, 104, 114, 143, 144, 165, 232

Trade: 19, 22, 25, 32, 33, 78, 80, 85, 103, 104-107, 184, 185, 191, 207, 208, 227, 229, 240, 250, 251, 255.

Undernourished: 1, 8, 71, 74, 86, 87, 103, 104, 113, 115, 212, 219, 232, 235, 236, 252

United Nations: 1, 15, 29, 30, 33, 37, 86, 114, 115, 165, 219

Value chains: 4, 22, 29, 31-33, 252.

Way forward: 11, 134, 159, 235, 250.